IN THE SUBURBS OF HISTORY

Modernist Visions of the Urban Periphery

In the 1960s, urban planners, architects, and city officials – socialist and capitalist alike – chose the urban periphery as the site to test out new ideas in modernist architecture and planning. *In the Suburbs of History* examines the outskirts of Prague and a bedroom suburb of Toronto as sites for this experimental urban development.

The book overcomes the divisions between East and West to reassemble the shared histories of modern architecture and urbanism as they shaped and reshaped the periphery. Drawing on archives, interviews, architectural journals, and site visits to the peripheries of Prague and Toronto, Steven Logan reveals the intertwined histories of capitalist and socialist urban planning.

From socialist utopias to the capitalist visions of the edge city, the history of the suburbs is not simply a history of competing urban forms; rather, it is a history of alternatives that advocated collective solutions over the dominant model of single-family home ownership and car-dominated spaces.

(Global Suburbanisms)

STEVEN LOGAN is an adjunct faculty member at the Institute of Communication, Culture, Information and Technology at the University of Toronto Mississauga.

GLOBAL SUBURBANISMS

Series Editor: Roger Keil, York University

Urbanization is at the core of the global economy today. Yet, crucially, suburbanization now dominates twenty-first-century urban development. This book series is the first to systematically take stock of worldwide developments in suburbanization and suburbanisms today. Drawing on methodological and analytical approaches from political economy, urban political ecology, and social and cultural geography, the series seeks to situate the complex processes of suburbanization as they pose challenges to policymakers, planners, and academics alike.

For a list of the books published in this series see p. 227.

STEVEN LOGAN

In the Suburbs of History

Modernist Visions of the Urban Periphery

UNIVERSITY OF TORONTO PRESS
Toronto Buffalo London

Toronto Buffalo London
utorontopress.com

ISBN 978-1-4875-0788-6 (cloth)
ISBN 978-1-4875-2543-9 (paper)
ISBN 978-1-4875-3715-9 (EPUB)
ISBN 978-1-4875-3714-2 (PDF)

Global Suburbanisms

Library and Archives Canada Cataloguing in Publication

Title: In the suburbs of history : modernist visions of the urban periphery / Steven Logan.
Names: Logan, Steven, 1974– author.
Series: Global suburbanisms.
Description: Series statement: Global suburbanisms | Includes bibliographical references and index.
Identifiers: Canadiana (print) 20200310054 | Canadiana (ebook) 20200310135 | ISBN 9781487525439 (paper) | ISBN 9781487507886 (cloth) | ISBN 9781487537159 (EPUB) | ISBN 9781487537142 (PDF)
Subjects: LCSH: Suburbs – Ontario – Toronto – History – 20th century. | LCSH: Suburbs – Czech Republic – Prague – History – 20th century. | LCSH: City planning – Ontario – Toronto – History – 20th century. | LCSH: City planning – Czech Republic – Prague – History – 20th century. | LCSH: National socialism and architecture – Ontario – Toronto. | LCSH: National socialism and architecture – Czech Republic – Prague.
Classification: LCC HT352.C32 T67 2020 | DDC 307.7609713/541 – dc23

This book has been published with the help of a grant from the Federation for the Humanities and Social Sciences, through the Awards to Scholarly Publications Program, using funds provided by the Social Sciences and Humanities Research Council of Canada.

University of Toronto Press acknowledges the financial assistance to its publishing program of the Canada Council for the Arts and the Ontario Arts Council, an agency of the Government of Ontario.

Canada Council for the Arts
Conseil des Arts du Canada

Funded by the Government of Canada
Financé par le gouvernement du Canada
Canada

Contents

Illustrations

Preface

"How did he wash up in the Czech Republic? I certainly wanted to know." So asked one of this book's anonymous reviewers. I took the comment in the jest and irony for which it was certainly intended, as the Czech Republic is a landlocked country. In his Czech history *The Coasts of Bohemia*, Derek Sayer reminds us that it was Shakespeare, in *The Winter's Tale*, that gave Bohemia a shoreline. Sayer writes that "Czechs are inclined to see Shakespeare's furnishing of their country with a coastline as typical example of foreigners' ignorance of their land."

If in more contemporary times there is a space that is not well understood outside of Czech circles it is the *sídliště*, as the Czech post-war socialist housing developments are known. My experiences with the sídliště began with my very first visit to the Czech Republic in 2000, at the Sídliště Novodvorská, where a Canadian friend was living with his Czech girlfriend. But it was not until I met my future partner Hana, followed by my move to Prague in 2003 so that we could live together, that my understanding of socialism and its housing completely changed. Spending our first year together in Letňany, another Prague housing development in the northern part of the city, our conversations often turned to our respective upbringings: mine, in a suburb of Toronto, and hers, in the new town of Havířov, just outside of the city of Ostrava. In these conversations, the seed of this comparative project began to germinate.

It was not that our worlds were similar, but it was as if they were connected through the looking glass. If I walked through the mirror, I could just as easily have found myself at a flower parade in Havířov, as she could have found herself at a birthday party at a McDonald's. Before I had a background in modernist architecture, I had found her descriptions of running out of her apartment into a nearby playground or the forest without the fear of traffic to be a suburban experience, minus the cars and the single-family houses. On the weekend, she had easy access to the mountains, where she travelled with her family – Havířov's proximity to the mountains offered a socialist reward for her parents' work in the factory and the mine. If my childhood family vacations were spent travelling to Florida and Disney World, then hers were the factory-sponsored trips to luxurious hotels that they would otherwise never have been able to afford. Granted, Havířov was its own new town, but it could only exist because of its proximity to the mining and steel-production centre of Czechoslovakia, Ostrava.

As I spent time in Havířov, and as our children grew to love the concrete apartment block in which their mother grew up, I began to question two long-held assumptions: that the suburb could mean something other than single-family homes and family cars, and that life under socialism, by no means perfect, had pockets of idyllic modernism. As I began to study modernist architecture, and especially its critics, I realized they had probably never been to a place like Havířov. Of course, this is not at all to idealize life there – it had and continues to have its share of difficulties – but the utopian thrust of the town can still be glimpsed on its streets and in its parks and forests. There were and are more ways of being suburban, and more ways of being modern.

This book could have easily been only about Czech modernist architecture and urbanism. But when I began to research the Toronto suburb of Willowdale as part of my work with the urban research collective LOT: Experiments in Urban Research, I uncovered a history that offered parallel grand visions for the urban periphery. This archival research was part of the Leona Drive Project – a site-specific art installation in six post-war houses curated by Janine Marchessault. The world opened up to me as the project's archivist prompted me to continue working on Willowdale in this book. In particular, it was the 1968 plan to redevelop Willowdale into a new suburban centre that caught my attention. Here was one of Canada's foremost architects – John C. Parkin – and one of the country's leading urbanists – Murray V. Jones – proposing to build a suburban version of the CN Tower. The project matched the sort of ambition that I would discover in the 1960s plans for Prague's socialist cities, all built on the urban periphery.

This book is an attempt to work through the commonalities and differences of the modern suburb. It is the culmination of a suburban life left in Toronto and returned to on the Prague periphery. Behind the research and the theoretical underpinnings is a born-and-raised suburbanite making sense of and bringing together these two parallel worlds.

Over the many years that this project has been in development, a number of people have helped shape it and offer ideas amidst the growing complexity of a comparative project. I would especially like to thank Jody Berland, Janine Marchessault, Stefan Kipfer, Robert Fishman, Pierre Filion, and several anonymous reviewers. Kimberley Elman Zarecor offered valuable and insightful suggestions on earlier drafts of this manuscript. My thanks also go to outgoing editor at University of Toronto Press Doug Hildebrand and incoming editor Jodi Lewchuk, copy editor Ryan Perks, associate managing editor Robin Studniberg, and indexer Ellen Hawman.

Over the years, the friendship provided by Anna Friz and Jason Rovito has been a much-needed constant in precarious times. They have offered directions, conversations, comfort, and revelry.

Thanks also go to Roger Keil, the editor of this book series, whose support and enthusiasm for my work has been unwavering. It was so fortuitous that I first met Roger as he was preparing the application for what would become the seven-year Global Suburbanisms project. I was very privileged to be a fellow traveller on this international network of scholars rethinking the suburb.

A treasure trove of accessible archival material from the Canadiana Department at the North York Public Library was an invaluable contribution to my Willowdale research, and in particular the shelf full of binders of historical material compiled by the North York Historical Society. In one of those binders, I stumbled upon a speech given by North York reeve James G. Service at the unveiling of the Willowdale redevelopment plan, which he boldly compared to the Weissenhof housing experiment. The Canadiana Department was closed when the library began renovations in 2016, its contents distributed to other library spaces. Serendipitous encounters with historical ephemera are rare, and the Canadiana Department, small and modest as it was, allowed for it. My thanks go to all those librarians who kept it going for so many years.

My research at the Canadian Centre for Architecture in Montreal was generously supported by the Landscape Architecture Canada Foundation.

This book, and in particular chapter 4, on South City, owes much to the people who helped shaped the spaces, both physical and symbolic, it describes, above all the architects Vítězslava Rothbauerová and Jiří Lasovský. I would also like to thank Jiří Sulženko, local chronicler Jiří Bartoň, Karel Maier, Luděk Sýkora, Jan Dostalík, Miroslava Fišarová and Jindřich Štreit, the documentary photographer with whom I happened to share a train compartment on a journey to Prague and who put me in touch with Jaromír Čejka; Čejka's photographs of South City in the 1980s have inspired and drawn many people to that place, including me. I would especially like to acknowledge the sociologist

Jiří Musil, who in the late stages of cancer, opened his home to me and gave his time so generously to talk about his involvement in urbanism. His life and work have influenced Czech scholars revisiting the socialist suburbs and I am so grateful I had the opportunity to hear him speak about his life.

The last words go to those who have been there from the start: My brother Jeremy and sister Lisa, my parents Beryl and Shelly. And to Hana and our two children, Asher and Nina, with love, I give you this book.

IN THE SUBURBS OF HISTORY

Modernist Visions of the Urban Periphery

1 Introduction: Crossing Divides

In August 1967, the newly built York University, in the Toronto suburb of North York, hosted a massive, 10-day conference on "metropolitan problems" that drew architects, politicians, and planners from major cities around the world, including delegates from a handful of socialist cites: Prague, Moscow, Leningrad, Belgrade, Warsaw, and Budapest. Much was made in the Canadian press of the visitors from Moscow, who claimed they had no problems because they had restricted population growth in their city and combated congestion by focusing on public transport and taxis rather than private cars. Jiří Hrůza, Prague's deputy city planner, was part of the five-person Czechoslovak delegation – he was one of Prague's most influential planners and theorists, who in his work on socialist urbanism and architecture drew on a plethora of examples from both the West and the Soviet Union. According to an article on the conference in the Prague daily newspaper *Večerní Praha*, Hrůza's slide presentation on the history of Prague urbanism proved so popular that he gave it a second time. Hrůza was probably sitting in the audience while North York mayor James Service, a champion of the modernist plan to redevelop Willowdale as North York's new downtown, welcomed the delegates. Hrůza also likely travelled to Montreal for Expo 67, which was going on at the same time. The Czechoslovak pavilion, most noted for its multimedia installations, included a display model of an experimental suburb in a section entitled "Conflicts." Another key element of Expo, and which still stands today, is architect Moshe Safdie's Habitat, a prefabricated apartment megastructure offering a garden in every unit. It was funded by the Central Mortgage and Housing Corporation (CMHC). The urban planner and theorist Humphrey Carver, who worked at CMHC and helped shape Canada's post-war suburbs, encouraged Safdie to pursue the project, and he also stayed there with his family.

At the conclusion of the York University conference, many promises were made for future international cooperation, with Toronto and Soviet planners hoping to take part in a professional exchange, a reflection of the cooperative atmosphere in which the conference occurred. That Hrůza, along with the deputy mayor of Prague and three other Czechoslovak attendees, came to Canada reflected the growing ease with which architects and planners travelled in the 1960s as part of the liberalizing atmosphere of the Prague Spring. Only a few months previously, Prague hosted the triennial International Union of Architects (UIA) conference under the theme "architecture and the human environment."

This spirit of the international cooperation provides the jumping-off point for *In the Suburbs of History*, which looks at the production of alternative forms of suburban space under socialism and capitalism in the Prague suburb of Jižní město (South City) and the Toronto suburb of Willowdale. This book makes connections and draws parallels with phenomena usually kept separate: socialist and capitalist cities, both real and envisioned. They were not diametrically opposed, but rather deeply connected because architects, urbanists, and theorists engaged a shared body of work on modernist urbanism.

This does not accord with the common-sense ways in which the histories of communism and capitalism are understood, separated by Winston Churchill's "iron curtain." In Europe, that divide was especially porous when it came to architecture and urbanism. For example, copies of architectural journals like *L'Architecture d'Aujourdhui* and *The Architectural Review*, along with journals from Scandinavian countries, were available to architects in the Eastern Bloc (Moravánszky et al., 2017, p. 8). In Prague, the editors of *Architektura ČSSR*, the main architectural journal in the country, sent issues to their French colleagues, and they would receive French journals in return, like *L'Architecture d'Aujourdhui*. In the mid-1960s, the borders were opened to travel and so architects and urbanists could journey to places like Montreal and Toronto as well as to other conferences and workshops, and they took study trips to learn from the West.

As the above examples show, the porous divide was not simply within Europe, but beyond as well, and nor was it simply one of exchange between countries. For example, Scandinavian architecture and urbanism was very influential in Czechoslovakia, particularly in the post-war period, and Canadian urban planners similarly drew on examples from Scandinavia in their own post-war architectural thinking. The Stockholm suburb of Vällingby, for example, was an important global influence across the Cold War divide. Even though socialist architects sought as their end goal the production of a distinctly socialist space, borrowing from the West was allowed if not encouraged, particularly in the wake of the Khrushchev Thaw, beginning in the 1950s. Even during the 1920s in the Soviet Union, architects and planners drew on a number of Western sources to such a point that "borrowings from the West ... were labelled as socialist by Soviet practitioners" (Bittner, 1998, p. 26). And in the fervour for socialist city building in the 1920s and '30s, many noted modernist planners and architects from the West, particularly Germany, worked on new city-building projects in the Soviet Union. One of them, Hans Blumenfeld, would end up in Toronto in the 1950s working on Metropolitan Toronto's first official plan.

In the Suburbs of History, then, challenges the divide between capitalist and socialist suburbs, offering a case for their comparability by looking at the shared histories of modernist urbanism upon which architects and urbanists drew as they attempted to rebuild, or build anew, the urban periphery. In doing so, this book makes four basic arguments. First, *In the Suburbs of History* challenges the dominant "white-picket-fence" view of the suburbs as neighbourhoods of owner-occupied, single-family houses with two cars in the garage. This is one particular history tied to dominant ideologies around home and car ownership. There were and still are many different models of life on the urban periphery. While the turn towards density and pedestrian-friendly environments have become the focus of twenty-first-century city urbanism, this book suggests that this is not a new phenomenon, but instead can be traced back to the debates in the 1960s. Willowdale and South City were to be key nodes in the urbanization of their respective cities. Although peripheral to the existing understandings of suburbs, they are two early and important examples of what theorists now refer to as post-suburbia (e.g., Phelps & Wu, 2011).

Second, in the histories of modernist urbanism and suburbanization, the 1960s represent a foundational moment for the formation of the global suburb and the history of global suburbanization – today's "edge cities" (Garreau, 1991) and "technoburbs" (Fishman, 1987) can be traced back to these debates. In both capitalist and socialist cities, architects, urbanists, and theorists drew on modernist urbanism as a way to reform the perceived deficiencies of their respective suburbs: the sprawling landscape of single-family houses synonymous with suburbia and the vertical sprawl of the prefabricated concrete apartment blocks that dominated the urban peripheries of the Eastern Bloc and the Soviet Union.

Third, the book challenges a monolithic understanding of modernist urbanism, which was in reality a diverse body of ideas. In its pejorative sense, modernist urbanism is usually understood as a set of design principles that rejects past urban forms and lifestyles in favour of entirely new ways of building, dwelling, and moving in urban space – old planning is obsolete and no longer applies. More often than not, these ideas were tested at the peripheries of cities. Critics of the heroic modernist urbanism associated with its most noted figures – Le Corbusier, Robert Moses, and Walter Gropius – believed that this approach necessarily leads to "urban dystopia" (Keil, 2018, p. 113).[1]

Reducing modernist urbanism to its most celebrated, and denigrated, figures has been a feature of urban theory, even among its most noted thinkers like Henri Lefebvre and Jane Jacobs. Although the plans for Willowdale in many ways adopted this heroic modernism, the inclusion of the socialist suburb of South City, and the wider conversations within the Eastern Bloc more generally, offer a far more diverse picture of modernism, one that does not reduce it simply to design gestures, particularly when it is connected to the movement for a socialist humanism in 1960s Czechoslovakia: addressing the inequalities inherent in capitalist urbanization by providing decent, affordable housing in a quality living environment. To grasp the full diversities of modernist urbanism, *In the Suburbs of History* connects modernism's design gestures to the ideological orientation of the respective suburbs: in Willowdale, the development was part of a larger discourse around automobility and regional planning in Toronto and the need for a network of urban subcentres to take pressure off the downtown core; South City, on the other hand, was part of the wider humanist-socialist project of reforming the existing housing on the urban periphery.

In connecting the socialist humanism of 1960s Czechoslovakia to the crisis of the socialist suburb (Hrůza, 1967a), the book also challenges the stereotype that socialism was simply an authoritarian model with an all-powerful state imposing its ideas on architects and inhabitants alike. Although the state certainly held an extremely powerful position, particularly in the implementation phases of building projects – South City is an exemplary case in this regard – the relationship is far more complex, with architects and urbanists accorded a degree of autonomy in the 1960s that allowed them to engage modernism from a socialist perspective even though they were also subject to the dictates of the state.

Finally, *In the Suburbs of History* engages and critiques the standard argument that "modernist urban landscapes were built to facilitate automobility and to discourage other forms of human movement" (Freund & Martin, 1993, p. 119). In one sense, twentieth-century modernist urbanism *was* about freeing urban space to facilitate the flow of motor vehicles, and the plans for both Willowdale and South City to varying degrees are landscapes shaped by automobile movement. Automobility's dominance in this period is not questioned. However, that these landscapes were planned to discourage other forms of movement reduces and simplifies the diverse ways in which modernism sought to also facilitate pedestrian movement and contribute to the creation of gathering spaces for people, with car-free centres, green spaces, and elaborate walkways. There was life between buildings, to quote the noted urbanist and modernist critic Jan Gehl, it just may not have been what downtown-minded planners had in mind. The relationship was a contradictory one, as urban plans both facilitated and restricted the spread of automobility. The history of the automobile is inseparable from the growth of the North American suburb, particularly in the post-war period, while the socialist suburbs are more often associated with public transportation even though they were designed with generous road systems. Using the car as one of the bases of comparison, along with the single-family house and the apartment block, *In the Suburbs of History* allows a more complex picture of suburbanism and its alternatives to emerge.

Modernist Urbanism

In *The Production of Space*, Henri Lefebvre ([1974] 1991) attempts to trace the moment when architects began to think about not just individual buildings, but the production of space as a whole, a space that could be replicated globally. Beginning in the 1920s, Lefebvre argues, the architects, artists, and engineers working with the Congrès International d'Architecture Moderne (CIAM) "developed a new conception, a global concept, of space" (p. 124). Roger Keil (2018) echoes Lefebvre's emphasis on the historical importance of CIAM and brings it out to the suburbs, suggesting that this new theory of space was for the most part implemented on the periphery, be it in completely new towns away from the existing dominant cities or housing developments on the peripheries of existing cities like the *sídliště*, as they are known in Czech, and the *Siedlung* in Germany.

The work of CIAM is central to the modernist production of space. CIAM was founded in 1928 on the idea that urban life could be reduced to four functions: dwelling (*habiter*), work (*produire*), recreation (*se delasser*), and transportation and communications (*circulation*) (Gold, 1997, p. 59). They would become the main subject of discussion at CIAM's fourth conference on the Functional City in 1933, which took place on a cruise ship that travelled from Marseilles to Athens. Ten years and many heated debates and discussions later, the Athens Charter was published – authored largely by Le Corbusier rather than the diverse collective of architects and urbanists who attended the congress – and it codified a language for talking about the city that became influential worldwide.

In exploring modernism as a diverse set of ideas, *In the Suburbs of History* also challenges the monolithic status accorded the Bauhaus and CIAM. Lefebvre ([1974] 1991) describes their "'programmatic' stance" as "tailor-made for the state – whether of the state-capitalist or the state-socialist variety" (p. 124). Like the global suburb, modernist urbanism was a diverse set of ideas – by the late 1950s, the breakaway CIAM group known as Team 10 was rejecting Athens Charter urbanism. Urban theorist Colin McFarlane's (2010) term "Corbusier's circulating modernist urbanisms" captures the way modernist urbanism held influence globally.[2] Urbanisms, in the plural, points to its multiple forms, although labelling them "Corbusier's" belies McFarlane's own critique of comparative urbanism, where there is an inclination to "compare with and learn from the 'usual suspects'" (2010, p. 728) – both people and places – rather than learning from the "worlding" of other more ordinary urban spaces and people (Cochrane, 2011, p. x; Robinson, 2006). Modernist ideas are far from monolithic and the tendency to equate modern urban planning with Le Corbusier obscures the range of approaches to modernism that have been taken (Deckker, 2000, p. 4). This applies to modernism on both sides of the political divide. More often than not, the socialist suburbs are in their design reduced to examples of Athens Charter urbanism with Le Corbusier as a stand-in for modernists from the West.[3] The relationship between socialist architects and modernist urbanism was far more complex than architects and urbanists simply reproducing Athens Charter urbanism throughout the Eastern Bloc.[4]

This has a double resonance, referring not only to urban theorists who examine New York, Paris, or London, rather than Toronto, Prague, or Warsaw, but to "actually existing comparative urbanisms" (Clarke, 2012): cities themselves and the way particular models of modernist urbanism are embraced over others. Many of the people and places that this book evokes will be familiar to readers, but many will not. The problem with sticking to a standard narrative of modernism governed by a few dominant figures and places is it often provides the basis for the outright dismissal of modernism, even though modernism, particularly when both its socialist and capitalist variants are explored, was as much an engine of diversity as an engine of homogenization. For example, although the modern architecture planning of Casablanca and Chandigarh bears the imprint of some of modernism's most noted architects, these cities were the "collective oeuvre" of not just planners and architects, but also sociologists, builders, construction workers, politicians, and international organizations (Avermaete & Casciato, 2014, p. 31). In Casablanca and Chandigarh, modernist urbanism was less "universal recipe" and more a result of "elaborate encounters, exchanges and cooperation between various transnational and local actors" (Avermaete & Casciato, 2014, p. 31).

There are two key aspects to modernist urbanism's influence on the suburb. The first lies with the superblock, a design principle that took on different forms under capitalism and socialism, but which in principle shared the desire to separate out cars and pedestrians, creating car-free pedestrian realms surrounded by arterial roads. This principle seemingly embraced the automobile by planning for its spread, while also keeping it in its place away from the communal spaces of neighbourhoods. Under capitalism and socialism, the degree of reliance on the automobile and the form of the superblock differed, both spatially and ideologically, especially given that the superblocks in the socialist suburb would be generously supported by public transportation, while the focus of the capitalist suburb was on enabling commutes to the city and jaunts to the countryside. The pedestrian spaces at the centre took on many forms going back to Ebenezer Howard's "cooperative quadrangles" in his proposals for garden cities, the grand green spaces of the socialist city, architect Victor Gruen's shopping malls, and the megastructures of the post-war new town.

The focus on the public spaces at the heart of the superblock points to the second defining feature of modernist urbanism: in place of the uniform horizontal and the vertical sprawl, the suburbs would be organized around key regional subcentres defined by their density and urbanity, and in many cases access to public transportation. The new suburban core or heart, as CIAM called it in their eighth congress on the "Heart of the City" in 1951, would bring public space, "new forms of community," and monumentality to both city and suburb. As with the superblock, the ways in which the suburban core would be envisioned depended on the political context and the ideology of the urbanists themselves, many of whom were aligned with CIAM, but who became increasingly critical of its rigid approach to urbanism. The subcentres, as they manifested in both the capitalist and the socialist context, are an ideal focus for this book's comparative approach because each in their own way sought to be representative of their respective society – although socialist architects and planners turned to the West for inspiration and influence, the end goal was the production of socialist space.

In the Suburbs of History explores how these core ideas of modernist urbanism were taken up and implemented in completely different ways in Toronto and Prague, two cities that were themselves peripheral to the histories of both modernist urbanism and suburbs. Although there are a number of works exploring modernist planning in Toronto and the suburbs – foremost among them Sewell (1993) – and Canadian suburbs more generally (Harris, 2004), the history of Willowdale's redevelopment into a downtown for the growing borough of North York has received scant attention. Similarly, the international influences on Toronto's suburbs are usually limited to the grand ideas of Ebenezer Howard and Le Corbusier, while the influence of figures like Humphrey Carver, whose key role in building and critiquing Canada's post-war suburb is discussed in chapter 5, is rarely considered.[5] Wider influences from CIAM are also lacking in the post-war history of Toronto's suburbs.

The urban planner Jaqueline Tyrwhitt, a key organizer of the 1951 CIAM congress – although peripheral to the modernist – male-dominated planning environment, played a significant role in post-war urban planning (Liscombe, 2007; Shoshkes, 2016). Tyrwhitt was a key translator of ideas, particularly between Europe and Canada (Darroch, 2008; Wigley, 2001). She would continue her focus on the core during her brief stay as a visiting professor at the University of Toronto from 1951 to 1954, where she was instrumental in developing a graduate program in urban planning. The anthropologist Edmund Carpenter, who worked with Tyrwhitt alongside Marshall McLuhan as part of the Explorations Group, said, "Jackie Tyrwhitt knew how to translate thought into reality. Never thanked, never credited, she helped change Toronto" (quoted in Shoshkes, 2016, pp. 175–6).

She left Toronto in 1954 for Harvard, where she became an assistant professor of city planning, and continued to work on ideas around urban cores. Although she was not involved in Willowdale's planning or the turn to urban subcentres, her thinking impacted a number of planners whose influence this book charts, like Carver and Blumenfeld.

Beginning with Toronto's first Master Plan in 1943, the major suburban boroughs around the city – North York, Etobicoke, and Scarborough – were the targets of population growth, with Willowdale identified as a growing population centre. The 1959 *Official Plan of the Metropolitan Toronto Area*, co-authored by Blumenfeld, did not focus on satellite towns, but on a "'finger' of development extending north from the city along Yonge Street," the city's major north–south artery (White, 2016, p. 86).[6] Although this draft plan was not officially adopted until 1980, it influenced planning in the 1960s and '70s (Sewell, 1993, p. 125).

Willowdale would be the largest node along this finger of development. When Metropolitan Toronto was formed in 1953, North York's population was 110,000, but by 1968, there were 425,000 people living in the borough, making it, at the time, the fourth-largest municipality in the country (Hart, 1968, pp. 258, 302). The authors of Willowdale's redevelopment plan noted that the greatest concentration of North York's residents was along the Yonge Street corridor and in Willowdale (Jones & Parkin, 1968b, p. 9). Willowdale was ready for its transformation. Overall, the three major suburban boroughs made up almost half of Metropolitan Toronto's population of 1.9 million (Metropolitan Toronto Planning Board, 1970, n.p.).

Willowdale's redevelopment was planned in large part to coincide with the extension of the subway into North York, which runs along Yonge Street; other subcentres within the city were already developing along the Yonge Street corridor at St. Clair and Eglinton. From King and Yonge, in the core of Toronto's downtown, to the cluster of Yonge Street suburbs in North York, Yonge Street was and is the city's key north–south axis. North York Centre, as it is now known, along with Scarborough Town Centre, were the two main suburban centres proposed in Metropolitan Toronto's official plan in 1980, even though development and plans in the area had been well underway since the early 1960s.

Scorn for the Suburb

That the suburbs needed reforming in the post-war period may be a topic for debate, but in the eyes of the modernist architects and urbanists this book discusses, the mass production of single-family homes in the West and apartment blocks in the East were cause for concern. Scorn for the suburb was particularly prevalent in the 1960s. In *The City in History* (1961), the urban historian Lewis Mumford made the provocative claim that the suburb, as such, was a thing of the past. Implicit in Mumford's claim was both a love for and a derision of the suburb. The mass-produced suburb evoked Mumford's ire, and *The City in History* is full of quotable passages displaying his antipathy for the mass-produced conformity of post-war North American suburbia. In a 1979 speech, Carver described himself as searching for a new vision for the suburbs, what he called "a 'post-suburbia' habitat."

In *Bourgeois Utopias*, the urban historian Robert Fishman (1987) writes that the post-war explosion of suburban house building that has generally become associated with suburbia was not the "culmination of the 200-year history of suburbia but rather its end" (p. 183). Although he suggests that suburbs can be understood quite broadly as "any kind of settlement at the periphery of a large city" (p. 5), Fishman's history focuses on the "middle-class suburb of privilege" (p. 5), the image that usually resonates in the popular imagination. When he declares the end of the suburb in the post-war period, he means the end of a specific kind of suburb, not the end to peripheral urban settlements.

More recently, Leigh Gallagher (2013) pronounced the end of the suburbs, suggesting that the American Dream is migrating to the cities, with the residents of "urban burbs" demanding that their suburbs look more like the traditional American and European city of old: more density, walkable, and close to public transport. This feels like familiar territory, simplifying the much more complex discussion that Dolores Hayden (1984) had already made in *Redesigning the American Dream*, where she points towards North America's unsustainable obsession with single-family homeownership. Hayden goes further and revisits some of modernism's more radical approaches to suburban planning, which included addressing unpaid women's work, providing cooperative living spaces, and affordable housing for single women. The desire for "urban burbs" did not just emerge in the twenty-first century out of nowhere and it was often deeply connected to social and political reform, not simply a free market decision. In this book, I trace this desire to the modernisms of the 1960s, which sought to reconcile the demand for automobiles with the demand for walkable, pedestrian-oriented spaces in the city and beyond, which CIAM made a priority in its congress on the core.

Both of the places studied in this book defy the conventions of suburbs. Willowdale's redevelopment explicitly sought to both preserve and overturn the common-sense definitions of the suburbs. The area around Yonge Street was filled with single-family homes, some of which were self-built in the 1910s and '20s on newly paved roads in what was largely a rural landscape. In the immediate post-war period, when rapid urbanization and population growth in North York took off, many cookie-cutter houses were mass-produced in the style of the Levittown suburbs in the US. The plans, in theory, would leave these neighbourhoods intact, while intensely redeveloping the area in around Yonge Street into a high-rise centre that was definitely not typical for the post-war suburb. Suburb and city would coexist in an entirely new kind of urban landscape.

The suburb is not usually associated with the massive apartment blocks built on the peripheries of the post-war socialist cities, however, if suburbs are understood generally as peripheral urban developments then these massive developments are a socialist version of the suburbs (Häussermann, 1996, p. 218; Hirt & Kovachev, 2015). Although these developments were often part of the city proper, like their counterparts to the west, they were generally located far from work, the city centre, cultural destinations like the cinema and theatre, and in some cases schools (Musil, 1985, p. 19). South City was to be the largest of Prague's post-war socialist suburban developments, built for 80,000 inhabitants in a city that, at the time the development was first proposed in 1964, had a population just over a million (Český statistický úřad). Unlike Willowdale, however, South City was to be an entirely new development on empty land, with only portions of the existing pre- and interwar housing stock to be preserved.

Czech geographers Luděk Sýkora and Ondřej Mulíček do not accept the label "suburb" for what is a "direct application of functionalist city planning principles" (2014, p. 136), which is why, in part, when these places are discussed they are almost invariably discussed in reference to the Athens Charter. The Czech sociologist Jiří Musil was one of the first researchers to undertake sociological research in the post-war suburbs through his seminal work *Lidé a sídliště* (People and Housing Estates) (1985) and the Výzkumny ústavu výstavby a architektury (Institute of Architecture and Town Planning, or VUVA, est. 1951). He argued that the charter was the strongest influence on the development on Czech housing developments, although he notes that the charter itself was only first published in Czech in 1964 (Musil, 1985, p. 32). Even if they look different from the nineteenth- or early twentieth-century city, they are still part of the city-building project. In spite of his unease with the suburban label, Sýkora suggests that the post-war developments *do* exhibit "typically suburban characteristics," such as their "peripheral location" and the exclusive focus on residences (Sýkora & Stanilov, 2014, pp. 259–60), although in South City's case, this was not necessarily intentional. Even the *chata*, or country house, to which residents flee to every weekend is a contender for the socialist suburb (Hirt & Kovachev, 2015, p. 177).[7]

In the 1960s, the socialist suburb came under increasing scrutiny, often through efforts to humanize the sídliště. Socialism with a human face, as the reform-minded president of Czechoslovakia Alexander Dubček called it, would produce new settlements that would radically differ from the mass-produced blocks of apartments that were criticized by inhabitants, architects, and sociologists in the Central and East European socialist countries: Poland, Hungary, Czechoslovakia, Yugoslavia, Estonia, and others (Moravánsky et al., 2017).[8] In Czechoslovakia, critique came both in the popular press and in architectural journals. A 1967 article in the daily newspaper *Večerní Praha* asks: What is life like in the sídliště? "The response of citizens was unequivocal: TERRIBLE." Of course, the sets of problems that each suburb was responding to were unique to the political context. Whereas the planners and architects in the West sought to control land speculation, in the Eastern Bloc, where land was de-commodified, the struggles were often between architects and the state building companies focused on mass-producing and constructing apartments as quickly as possible.

Instead of announcing, lamenting, or celebrating the end of the suburbs, *In the Suburbs of History* identifies the modern suburb as above all a historical process of dynamic growth and change on the periphery of cities. The suburb, like Henri Lefebvre's definition of the production of space, is a process rather than a static thing. The numerous critiques of the suburb point also to its endless mutability, and as such, there can be no one definitive definition of what the suburb is. This is compounded by the fact that places that were once suburban – a development on the urban periphery – are now urban, in the sense that there are many more developments further afield. Yet at the same time, these now-urban places still exhibit their suburban characteristics. The work presented here widens the definition of the suburb not geographically, but ideologically and politically, looking at South City, the largest settlement on the periphery of socialist Prague, and comparing it to Willowdale, a suburb turned into an urban subcentre.

Automobility

In his introduction to the edited collection *The Socialist Car*, Soviet historian Lewis Siegelbaum (2011) draws on the concept of "entangled modernities" (David-Fox, 2006) to better understand the ways in which socialist and capitalist approaches to automobility were mutually implicating. Here, automobility means not just the car itself, but the whole system that contributes to the automobile's existence, the effects of which are felt most profoundly in the suburbs: an ideology of freedom and individualism realized through the power and speed of the automobile; the actual experience of driving a car and the feelings of "placelessness" that emerge in car-dominated suburban landscapes; and "auto space" and the transformation of both suburban landscapes and domestic family life (Freund & Martin, 1993).[9]

If socialism is not usually associated with suburbs, at least in the classical definition of the term, the freedom of the open road and the automobile is even less so; Siegelbaum notes that the ideological aspect of automobility would have given socialist ideologues the most problems. And more generally, in Freund and Martin's understanding, automobility shapes and moulds the inhabitant into a car consumer, whereas under socialism these ideals and their consequences – congestion, car-clogged streets, lack of public life – were seen as very much *un*-socialist. Socialist states focused on public transportation and during his tenure as Soviet president, Khrushchev was "ideologically hostile" to private car ownership (Siegelbaum, 2008, p. 84). However, by 1964, change was already afoot when Khrushchev was replaced by Leonid Brezhnev, a "fervent automobilist" who it is said had a fleet of 12 foreign-made luxury cars (Siegelbaum, 2008, p. 241). Khrushchev's "austere socialism" was out of touch with the "consumer socialism" and the renewed focus on leisure that became dominant in the Eastern Bloc in the mid-1960s and '70s (Siegelbaum, 2011, p. 4).[10] In Prague's urbanism and architecture of the same period, there is a clear tension between the attempts to regulate the car's spread by creating pedestrian spaces and promoting public transportation in the interests of collective socialist life, on the one hand, and the ample road and parking spaces planned as part of an extensive and at the time oversized road system, on the other. Eastern Bloc countries sought to both accommodate the automobile and to cater to pedestrians and eliminate the congestion, pollution, and traffic jams synonymous with capitalist automobility run amok. These entangled modernities are central to the place of the automobile in the urban designs of Belgrade (Le Normand, 2011), Marzahn, East Berlin (Rubin, 2011), and city centres in the GDR and the USSR (Beyer, 2011).

The redevelopment of Willowdale in the late 1960s occured at a key moment in the history of twentieth-century automobility in North America, coinciding with Jane Jacobs's move to Toronto in 1968, the same year in which the plan to redevelop Willowdale was first made public. In envisioning North York's centre, the plan counted on not just the future subway, but the expressways built – especially the east–west Highway 401, just to the south of Willowdale – and in the pipeline. One of the key proposed highways was the Spadina Expressway, originating west of Willowdale and running south–east into the downtown core. The Stop Spadina Save Our City Coordinating Committee, of which Jacobs was a part, fought to keep this expressway out of downtown Toronto, and in 1971 the Province of Ontario canceled the project; it lead to a wave of reform-minded planning in Toronto (Sewell, 1993, pp. 174–98). The expressway would terminate in North York at Lawrence Avenue instead of in downtown Toronto.

North York's politicians were very much in support of the expressway, and in the wake of its cancellation the "Go Spadina" campaign, spearheaded by North York councillor Esther Shiner, fought to finish building the expressway further south to Eglinton Avenue. Although the City opposed the extension into the city, which would dump motorists onto Eglinton Avenue, it was approved in 1976. That year, in a *Toronto Star* article on urban expressways and the Spadina Expressway in particular, Mayor David Crombie said that "We are now back to 1971 and it's like going back to 1939 to us. We thought the war was over."[11]

Although this book is not about the Spadina Expressway's fascinating history, Willowdale's redevelopment straddles the anti-expressway debate in the late 1960s and the attempts in the 1970s to reform the kinds of modernism that privileged automobility above else (like GM's Futurama exhibition at the 1939 New York World's Fair, to which Crombie was ostensibly referring).

Like Tyrwhitt, Jacobs may not have directly influenced the planning of Willowdale, but the anti-expressway sentiment and the decision to cancel the Spadina Expressway certainly reverberated in Willowdale, when the redevelopment was under discussion. The discussions over urban highways, North American suburbs, and new socialist living spaces, can in part be distilled into the positions that Tyrwhitt and Jacobs occupied: a focus on classic cities in Jacobs's work, and Tyrwhitt's focus on a modernist urbanism more attuned to the region. Tyrwhitt draws out this tension in her review of *Death and Life of Great American Cities*. Although Tyrwhitt calls the book "magnificently provocative and stimulating" in her opening paragraph, in a subsequent sentence she calls it "prejudiced, arrogant, inaccurate and overstated" (1962, p. 197).

Tyrwhitt believed that Jacobs's reference points for modernism – Ebenezer Howard and Le Corbusier – and the ideas she discussed related more to the modernist urbanism of the 1930s than the 1960s. Although Tyrwhitt downplays the persistence of Athens Charter–type modernist thinking well into the 1960s – the strict separation of functions – more generally she writes that the ideas of contemporary planners and architects in 1962 were "exactly that of Jane Jacobs: more urbanity, higher densities, greater liveliness, let the cities live" (1962, p. 198).

As the automobile and the low-density horizontal sprawl and energy consumption with which it is associated comes under increasing scrutiny in the present, we would do well to return to these moments when architects and urban planners sought to respond to these problems by attempting to bring centrality and urbanity to the mass-produced suburban environments. In the 1960s, when Jacobs's and Tyrwhitt's ideas held sway globally, there was serious debate over how to respond to the growing spread of automobility, which often lead to radical proposals for pedestrian-oriented spaces, not just downtown, but on the periphery.

The late 1960s was an important moment of global cooperation and struggle over what cities and their suburbs should look like. Across political divides, Toronto and Prague were key participants in this conversation.

History as Itinerary

The title of this book implies a place that diverts from the main road, not necessarily a cul-de-sac or a dead end, but one that branches from the dominant understandings of the suburbs and as such reshapes what gets to be a part of suburban history. Urban history can be spatialized – some places are neglected, relegated to the periphery, while other bright-lights, big-city topics are privileged.[12] It also suggests that some sort of navigation will be necessary; I have used the metaphor of the itinerary, which implies both a plan for travel as well an account of that journey. *In the Suburbs of History* is both. It is also an attempt to recover the full modernism of these places, as well as their ideological and political content. Urbanists and architects may have drawn on a shared history of modernism, but the socialist suburbs were still, first and foremost, socialist.

Comparing and taking seriously the suburbs in different locales and under different political conditions contributes to renewed discussion in urban studies on suburbanisms and suburbanization and thinking with rather than against the suburbs.[13] The inclusion of a socialist city also upsets common-sense, and ultimately restrictive, understandings of the suburb. Sonia Hirt's (2012) work on Sofia, Bulgaria has been important in this regard, as she suggests that the mass-produced apartment blocks built on the periphery of socialist cities are a socialist version of the suburb (p. 183). In this way, the classical picture of suburbanization is a departure point, from which this book charts various directions.

As the title suggests, this book is concerned with visions for the urban periphery. As such, it draws on a wealth of archival sources, both personal and institutional, old urban plans and models, and unpublished materials. In the Czech context, I supplemented my research with semi-structured interviews where necessary, given that I was working in a historical context with which I was on the whole unfamiliar. However, the principle reason for the interviews in the South City case was the opportunity to interview those directly involved with its conception: the architects Jiří Lasovský and Vítězslava Rothbauerová, as well as Jiří Musil, who in addition to his sociological work, was a consultant on South City and the experimental suburb Etarea. These interviews helped provide context and were therefore key to navigating a complex political, linguistic, and cultural terrain. In contrast, I grew up in Willowdale and witnessed first-hand the changes in the area, and it was that experience that proved invaluable to my involvement in the Leona Drive Project, a site-specific art project that took place on a cul-de-sac in Willowdale in and around six small post-war houses in 2009. My work as the project's archivist took me to many different libraries and archives, and it was during that research that I happened upon the redevelopment plan for Willowdale. Archival research also forms a significant part of the chapter on Humphrey Carver, whose archive at the Canadian Centre for Architecture, which inludes a wealth of unpublished material, offers a significant overview of his contributions.

As the research progressed, my understanding of a monolithic modernism so readily dismissed by urban activists, planners, and theorists from the 1960s on was increasingly tempered by a far more complex understanding, mediated in large part through my encounters with the left modernism of Karel Teige, one of the central figures in the Czechoslovak avant-garde during the 1920s and '30s, and a key intellectual translator between East and West. The connections between these two worlds became significant following the Bolshevik revolution in 1917, with architects and planners looking to the Soviet Union as the place in which modernity could emerge "shaped strictly on principles of functionality and equality, unsullied by private interest, sentiment, privilege or profit" (Cook et al., 2014, p. 809).

At the same time, the approach to modern architecture specifically is not in the form of a history of individual buildings, but rather architecture as an environment produced through a host of social, cultural, political, technological, and ideological factors. Cataloguing and noting individual buildings is difficult anyway because, as the examples of Willowdale and South City show, what *didn't* get built is as significant as what did – visions are rarely implemented as planned, if implemented at all. It is an architecture of the what-might-have-been or what-should-have-been (depending to whom one talks). It is difficult to find the appropriate metaphor to illustrate a layering of space in which buildings never did appear. In semiotics, what is not said is as important as what actually is. I adopt a similar approach to the built form: what did not get built is also important to the semantics of space. "Meaning is fluid," writes urban historian Maiken Umbach (2006), with "traditions invented, remembered, half-forgotten; identities tried out, and half-discarded; futures imagined, planned, defended, half-abandoned" (p. 14). Keil (2018) calls these places "complex palimpsests of tumultuous histories" (p. 127). My methodology consists of uncovering those palimpsests in order to put the built and the unbuilt into dialogue, my return visits to both Willowdale and South City guided by a desire to see if this particular detail was implemented, to navigate a road that had been thirty years in the making.

Navigating the built and the unbuilt, urban space and the space of models, plans, and architectural journals also speaks to modernism's complex history. In her book on architecture and mass media, Beatriz Colomina remarks on a parallel between Le Corbusier's buildings and the many historical studies that delve into his vast archive: "They are less about enclosure than about the entanglement of inside and outside, less about a traditional interior than about following an itinerary (no matter how many times redrawn, no matter how nonlinear)" (1994, p. 11). Modern architecture and the many modern technologies that emerged at the same time collapse the boundaries between inside and outside, between city and country, and between city and suburb. This book travels between places and between disciplines, seeing the global suburb as in motion in a process of building and rebuilding.

2 Looking for the Antithesis of the Suburb

If the history of modernist urbanism is viewed as an itinerary, then the ideal starting points for this transatlantic comparison are the publication of Ebenezer Howard's *To-morrow: A Peaceful Path to Real Reform* in 1898 (the book was republished in 1902 as *Garden Cities of To-Morrow*) and the founding of the Congrès International d'Architecture Moderne in 1928. Howard's vision for networks of decentralized cities situated amidst greenery and close to jobs influenced CIAM architects, particularly one of the group's founders, Ernst May, who as a city architect working in Frankfurt am Main helped transform the city's outskirts. May claimed that Howard's garden cities for 30,000 inhabitants were the first step in the evolution of the green city, culminating in the suburban *trabantenstadt* (satellite city). And, as Konstanze Sylva Domhardt's (2012) historical account of the connection between CIAM and Howard shows, there was much cross-over in ideas. The garden city's trajectory leads through CIAM and the satellite cities, suburbs, and new towns of Eastern Europe and the Soviet Union and towards North America, and the garden suburbs.

In his introduction to the 1946 edition of *Garden Cities of To-Morrow*, Lewis Mumford writes that Howard's garden city was "*the antithesis of a suburb: not a more rural retreat, but a more integrated foundation for an effective urban life*" (p. 35). The aim of this chapter is not to debate the merits of Howard's ideas or his legacy, but to take Mumford's claim as a call to examine the alternatives to the standard narratives of the suburb and modernism.

How did modern planners, architects, and theorists envision the alternatives to the dominant model of the suburbs? And what were those dominant models to which they were responding? There is a tension between monumentality and humanism, single-family homes and apartments, collective and private spaces at the core of the diverse suburbanisms of the twentieth century, both real and imagined. If the image of the suburb as a sea of single-family houses dominated by the automobile and consumerism has become the dominant suburbanism, the history presented here suggests a vibrant debate in different parts of the world on how to envision and build on the urban periphery.

Beyond the Athens Charter

CIAM's Athens Charter, developed in 1933 at the fourth CIAM congress but published only in 1943, has had a particularly powerful influence on urbanism both in the East and the West. By the 1960s, writes architectural critic Reyner Banham (1976), the Athens Charter had become "graven on the consciousness" of the architectural profession (p. 201). Banham notes that the sheer power the Athens Charter held over architects was due in part to the belief that it came solely from Le Corbusier (p. 201), which was not unfounded, as he wrote the final, published document. The charter, along with its author, was a frequent influence on the planning of the post-war socialist city, be it Marzahn in East Berlin (Rubin, 2011), socialist Sofia (Hirt, 2012), or New Belgrade (Le Normand, 2011), and the basis for the "tower in the park" neighbourhoods and housing projects in North America. One of the hallmarks of the Athens Charter that figured prominently in both modern urbanism and suburban planning was the separation of land uses and in particular the separation of car and pedestrian traffic. Giving people their own spaces, separated from car traffic, and thus off the streets, runs through the gamut of twentieth-century modernist urbanism. How this separation was imagined and achieved varied over time and space. The attempts to create pedestrian spaces in particular had a larger political aspect that infused the socialist approaches to urban planning, but also the struggle in both accommodating and resisting the spread of the car.

Architects and urbanists universally shared the idea that pedestrians and cars were to be strictly separated – in the age of the automobile, the traditional social space of the street, where transport and other functions coexisted, hampered efficient circulation for cars and placed pedestrians in harm's way. Modernist urbanism was united in its rejection of the traditional street where different modes of movement intermingled, where shops spilled out onto sidewalks, and where people lived above it all, looking down at the seeming chaos below. The charter called for parallel networks of footpaths for "slow-moving pedestrians" and a "network of fast roads" for cars. In his commentary on the charter, Le Corbusier writes that the sidewalks of the pre-automobile city are "absurdly ineffectual" because the new speeds have "introduced a real menace of death into the streets" (Le Corbusier, 1973, p. 64) and so thesis 62 requires that: "the pedestrian must be able to follow other paths than the automobile network" – no other change would bring about such a "fresher or more fertile era of urbanism" (p. 84). The Athens Charter demanded the purification of the street through the separation of pedestrians from vehicles, and streets themselves defined by the purpose they served (p. 85).

The separation of functions and wide streets dedicated to vehicular traffic, however, did not begin with the Athens Charter, which has a much longer history that Mumford (1961) traces as far back as to the "uniform oversized street" of the baroque city (pp. 367–71). Nor were CIAM architects and urbanists the first twentieth-century modernists to advocate this kind of planning. In Daniel Burnham and Edward Bennett's 1909 *Plan of Chicago*, the separation of cars and pedestrians achieved special prominence.[1] The plan called for separated traffic in the interests of economic efficiency and beauty. Like the CIAM values that would later come to dominate modernist planning, Burnham and Bennett identified three key problems with the street system: "disorder, inefficiency, and ugliness." The plan was first and foremost about making the city more amenable to business and raising property values, and it did so through a three-tiered street system that emphasized practicality and beauty, the hallmarks of which were boulevards combining street and park. The huge increase in automobile ownership in the ensuing two decades left some of the key aspects of the plan on the drawing board. Although Burnham emphasized both "beauty and practicality in the street system," by the end of the 1920s, the Chicago Planning Commission claimed "a near-exclusive focus on practicality, which ... had come to mean the narrow goal of accommodating automobiles" (Kling, 2013, p. 247). This turn of events became common among ambitious twentieth-century planners who sought to contain the spread of the automobile only to be later thwarted by its massive growth.

The separation of functions in the interests of efficient circulation was also a key feature of modernist architecture and urbanism in Prague, particularly in the interwar years. After the establishment of the Czechoslovak state in 1918, the Greater Prague Act of 1921 annexed the surrounding inner suburbs and villages and towns, raising Prague's population to 700,000 (Švácha, 1995, pp. 147–8). By the late 1920s "functionalist urbanism" and functionalist architecture had become increasingly important in Prague, with its focus on transportation and the division of the city into zones for working, industry, shopping, and dwelling (Švácha, 1995, pp. 164–5). Although there was not the same number of cars as in Canada or the United States, functionalist urbanism "paid unprecedented attention" to transport problems because automobile use was on the rise and because the separation of functions had exacerbated commuting problems for many inhabitants (Švácha, 1995, p. 166).

The Athens Charter, and the CIAM urbanism it represented, was not only about separation, but, in paradoxical fashion, the unity of the production of space as well. Henri Lefebvre in *The Production of Space* ([1974] 1991) attributed such importance to the work of architects and urbanists in the 1920s because he argued that they had worked out a theory for producing space as a whole: be it cities, districts, or neighbourhoods. In the global spread of space defined by separations, Lefebvre located the contradiction between unity and fragmentation. At the moment when architects and planners were claiming a new kind of space, they were, unwittingly or not, facilitating the tearing apart of the urban fabric, breaking it into pieces – the street, the house, the city itself – and reconnecting them again (badly, according to Lefebvre).

Lefebvre's critique calls out the remaking of space for the automobile and the separation of functions enshrined in the Athens Charter and implemented in the post-war modernization of both the socialist and the capitalist economies (Stanek, 2011, p. 148). Lefebvre is unequivocal in his assessment of the historic importance of CIAM's architects and planners: far from being revolutionary, they facilitated "the worldwide, homogeneous and monotonous architecture of the state, whether capitalist or socialist" ([1974] 1991, p. 126).

Although Lefebvre gestures to the "creative effervescence" of the early twentieth century in *The Production of Space*, he largely neglects both the internal struggles in CIAM in the interwar period and the post-war critiques of the Athens Charter; his discussion of the history of the production of space in the 1920s is limited to the familiar figures on the landscape, like Le Corbusier and Walter Gropius. The architectural theorist Łukasz Stanek has worked extensively with Lefebvre's writings, making just this point: even though these struggles and critiques aligned with Lefebvre's own critical writings on the city, he did not discuss them (2011, pp. x, 83).

The Athens Charter was by no means a jointly authored document. Ernest Weissman, a key member of the Yugoslav CIAM group at the Athens CIAM congress, was not in agreement with Le Corbusier's version and published his own version of events entitled "We had another charter" (Avermaete & Casciato, 2014, p. 61). The fissures and fractures within CIAM often centred around the relationships between the socialist-leaning members of CIAM, like Ernst May, Hannes Meyer,

Mart Stam, and Karel Teige on the one side, and Le Corbusier, Walter Gropius, and Sigfried Giedion, on the other, who believed that the four functions of the Athens Charter applied to all modern cities, socialist, capitalist, or otherwise (Mumford, 2009, p. 244). The fourth congress, where the Athens Charter was developed, was originally to be held in Moscow, and the proposals for that congress also were divided along political lines (Mumford, 2009, p. 245).

The conflicts within CIAM can be best understood in the context of relations between East and West. In his work on the relationship between CIAM and Soviet urbanism, architectural historian Eric Mumford delineates two periods: 1928 to 1933, and, in far more broad strokes, 1934 to 1959 (the last year in which the CIAM name was used). The early period, from CIAM's founding in 1928 to the 1933 congress on the Functional City, was an especially political moment in CIAM's history in that socialist ideas around architecture and urbanism played a key role in these early congresses, as the following chapter's discussion of Karel Teige's key writings on architecture and urbanism shows. Of course, the meeting point of socialist and modernist ideas did not end in 1959 with the dissolving of CIAM, but continued through the 1960s to the 1980s.

The Czechs were at the centre of these discussions, and in particular the meetings in 1929 on the *Existenzminimum* and in 1930 on *Rationelle Bebauungsweisen* (rational building methods). The Existenzminimum – small-sized dwellings affordable to the working classes and usually situated on the edge of cities – was more than just an architectural model (the somewhat unappealing combination of *Existenz* and *minimum* also conjures up all sorts of dystopian images): it was an approach to architecture and housing that was inseparable from social and economic problems, a key aspect raised by Teige and the Czech delegation to the 1930 CIAM congress in Brussels. Teige reported on the housing problem from the conference, where he emphasized that not only was the housing problem a socio-economic one, but it also concerned urban planning as a whole in addition to the building of individual houses (Teige, [1931] 1987, p. 150). This was Ernst May's approach as he headed the New Frankfurt initiative from 1926 to 1933, which saw 15,000 dwellings built in 14 new suburban settlements (Henderson, 2013). May had previously worked with garden city architect Raymond Unwin for two years on the Hampstead Garden Suburb in England. May was particularly inspired not simply by Unwin's traditional single-family houses, but by the formation of new settlements that offered good housing – both houses and apartments – as well as social infrastructure (Henderson, 2013, p. 17). In 1925, May attended the International Conference of

Figure 2.1. Layout of the Rabenhof apartment block in Vienna. Source: Steven Logan

the Regional Planning Association in New York, with Ebenezer Howard in attendance as well.

Although social housing in the 1920s usually found its place on the urban periphery, this was not always the case. The "Red Vienna" housing project (1924–32) explicitly rejected suburban garden cities for various political and economic reasons, and so for the most part the interventions took place within the already built-up city. Architectural historian Eve Blau uses the term "superblock" to refer to the *Gemeindebauten* (municipal apartment blocks) that "bridged streets and spanned several city blocks to create 'superblocks'" (1999, p. 287). The superblock did not just make one big block, but united multiple blocks as existing streets were incorporated *within* the superblock, contradicting the North American approach, which was to keep the superblock free of all automobile through traffic. In doing so, the *Gemindebauten* reverses the "traditional relationship between inside and outside, building and street" (Blau, 1999, p. 289). One such Gemeindebauten combines pedestrian and car separation with streets passing through the block: there is a monumental archway under which cars can pass, while other

Figure 2.2. Street in the Rabenhof apartment block in Vienna. Source: Steven Logan

pre-existing streets were closed to traffic and pedestrianized (see figures 2.1 and 2.2). The results, writes Blau, "blur the boundary between inside and outside," creating an environment that is both "intimate" and "idiosyncratic" (1999, p. 298).

The Building Superblocks of the Modern Suburb

This detour through the Vienna housing example offers a useful counter-image to the North American superblock. If the focus of CIAM and its left-leaning members was on collective spaces, social change, and economical dwellings, with May drawing on his work with the garden city, similar foundations informed the approach to building up the periphery in North America, but with the focus on the single-family home. The separation of cars and pedestrians was a key element in early North American suburban planning, which achieved wide recognition with Clarence Stein and Henry Wright's plan for Radburn, New Jersey, in early 1928. If the separation of cars and pedestrians was one of the key tenets of the Athens Charter, the garden city as a suburban building type was central to the formation of the production of modern suburban space, generated largely by North American planners influenced by European garden city ideals. They did so through the design of superblocks, arguably the suburban companion to the Athens Charter's focus on the separation of cars and pedestrians. The superblock and its corollary, the neighbourhood unit, are central to suburban development and modernist urbanism, and the post-war new town movement (Wakeman, 2016).

The story of the superblock usually begins with Stein and Wright's plans for both Radburn and the earlier Sunnyside Gardens in Queens, New York, but as with so many of the planning ideas this book discusses, the superblock was not confined geographically – indeed, it became a feature of modernist urbanism in places as diverse as Chandigarh, Casablanca, Vällingby, and Brasília.[2] Sven Markelius, one of the founders of CIAM and a key planner of Vällingby, cited the Radburn plan as an important influence – he visited Radburn while Stein visited Stockholm five times between 1949 and 1962 (Cook, 2018, p. 351). Of his 1949 visit, Stein remarked that the Stockholm city plan, which included Vällingby, "will be derived in large part from the Radburn Plan" (1949, p. 229). This is all the more remarkable given that Radburn was incomplete, unfinished because of the onset of the Depression; only two superblocks were built in the end.

The superblock has a modest beginning, at least in Stein's telling. It began with Henry Wright's experience of the semi-public space of an interior courtyard in Ireland in 1902: "I learned then that the comforts and privacy of family life are … to be found … in a house that judiciously relates living space to open space, the open space … being capable of enjoyment by many as well as by few" (quoted in Stein, [1957] 1966, p. 48). Wright was also taken with the way in which the courtyard was connected to and separated from the street outside it, leading to a luscious, semi-public garden, even though the central image of Wright's description is the "comforts and privacy of family life" in a house. Dolores Hayden (1984) writes that the models for this "neighbourhood strategy" were shared green spaces, courtyards, and the medieval cloister. Here, the home is not simply a stand-alone unit, but is connected to "semi-private, semi-public, and public spaces, and linked to appropriate scale and economic institutions assuring the continuity of human activity in these spaces" (Hayden, 1984, p. 124).

The "Radburn Idea," as Stein called it, was not just a way of replicating the European courtyard, it was an explicit response to the spread of the automobile, an answer to the "enigma" posed by Stein: "'How to live with the auto,' or … 'How to live in spite of it'" (Stein, 1949, p. 41). As Le Corbusier warned of the dangers of the corridor street for pedestrians, Stein (1949) called for an alternative to the standard gridiron layout typical of most North American cities of the time:

> [In 1928] the flood of motors had already made the gridiron street pattern, which had formed the framework for urban real estate for over a century, as obsolete as a fortified town wall. Pedestrians risked a dangerous motor street crossing 20 times a mile. The roadbed was the children's main play space. Every year there were more Americans killed or injured in automobile accidents than the total of American war casualties in any year. (p. 223)

Stein's Radburn Idea is significant for two main reasons: the strict separation of cars from pedestrians, and the houses turned away from the street and grouped around a common green, through which Radburn residents could access a system of pedestrian and bicycle paths to get to school or shops without ever having to cross a traffic artery (see figures 2.3 and 2.4). Radburn did not give the street back to children, but rather provided gathering spaces elsewhere. Essentially, the houses had two front doors; in this way the back became the front as much as the front became the back. A grouping of cul-de-sacs would together make up a Radburn superblock, with the courtyard replaced by a communal green space. A cul-de-sac makes no sense on its own, but only as part of the larger production of the superblock.

Although Radburn is often seen as the quintessential suburban model, it took inspiration from one of the most iconic urban spaces: Central Park in New York City, and Frederick Law Olmstead and Calvert Vaux's plan for separating out traffic, such that people on foot would never meet wheeled traffic (this was in 1851, a few decades before the automobile would make its appearance on city streets). Thus, Radburn was a Central Park writ large, and like the Central Park envisioned by Olmstead and Vaux, it was to stand as the antithesis of the surrounding neighbourhood, a respite in the midst of the bustling city.[3] Radburn's parks are an essential feature of the plan, as much as its rejection of the street as a social space.

In Wright and Stein's Radburn plan, the house symbolically turns its back on the noxious, crowded street, and towards the garden and the park. This gesture affirmed the divide between the city and the suburb

Figure 2.3. Houses turned away from the street towards a park in Radburn. Source: Library of Congress, Prints & Photographs Division, FSA/OWI Collection, LC-USF34–000620-D

and between the house and the street. To turn one's back on someone or something is a gesture of defiance and rejection. It reflects above all the contradiction between separation and unity that marks modern space, and which also signalled the rejection of the street in favour of communal green spaces. Through landscape design and urban planning, the house was to be separated from the street and the suburb separated from the city, yet at the same time, the expanding use of cars would prompt designs to unite house, suburb, and the surrounding region. Although the space encouraged people to walk and ride their bicycles, Radburn's location close to the highway and the newly built George Washington Bridge made it worthy of its title as the "town for the motor age."

Although it did not rewrite the existing grid network of streets, the housing in Stein and Wright's earlier Sunnyside Gardens (1924) was one of the first to create a separate network of pedestrian walkways as well as communal gardens at the heart of a superblock. Wright and Stein managed to create blocks of houses and apartment buildings

Figure 2.4. The pedestrian underpass for movement between the superblocks. Source: Library of Congress, Prints & Photographs Division, FSA/OWI Collection, LC-USF34–000673-D

that emphasized the inner courtyards and greens over the street, even though the tree-lined streets function much the way any of their urban counterparts would (see figure 2.5). Although Stein and Wright were attempting to import Howard's garden city into North America, they were equally influenced by the early colonial settlement Nieue Amsterdam (New York) in 1660, in which settlers built their homes on the periphery of a block with a garden in the middle (Stein, [1957] 1966, p. 44), not unlike some of the very early medieval cities that Mumford describes in *The City in History*: "houses … were usually built in continuous rows around the perimeter of their rear gardens; sometimes in large blocks they formed inner courts, with a private green, reached through a single gateway on the street" (1961, p. 282). A large apartment complex, the Phipps Garden Apartments, built in 1930, also includes a lush courtyard through which residents enter the building (see figure 2.6). The apartments were funded by the Society of Phipps Housing (initiated by steel industrialist Henry Phipps). The society funded housing for the working classes, but in Sunnyside's case the inhabitants would

Figure 2.5. Pathway through the common gardens of the Sunnyside Gardens superblock. Source: Steven Logan

be "white-collar clerical workers," who in Stein's thinking need "more spacious rooms and courts" (1949, p. 259).

These "demonstration projects" came out of the planning and housing advocates of the Regional Planning Association of America (RPAA), whose members included Lewis Mumford along with Stein and Wright (Hayden, 2009, p. 125). The RPAA understood neighbourhoods as "those that housed working families and preserved pedestrian access to schools, parks, shops, and transit, while accommodating the car" (Hayden, 2009, p. 126). All of the developments combined single- and multi-family housing. Hayden suggests that the area for social activity for city-minded people is often assumed to be on the street, but this is not always appropriate in the built spaces of the suburbs. Although Radburn was an attempt to import Ebenezer Howard's garden city into the North American context, the community was not surrounded by a green belt, as Howard had stipulated. Instead, the focus became the "peaceful green" at the heart of the superblock (Stein, 1949, p. 67). Hayden explores the social potential of "the centre of the block," a common village green in which, like in Radburn, the block

Figure 2.6. Courtyard in the Phipps Garden Apartments in Sunnyside Gardens. Source: Steven Logan

is turned inside out away from the car, simultaneously a private and public space (1984, p. 187). This common green can serve any number of purposes, such as food growing or daycare facilities, but as the Radburn plan was adopted and copied this important communal space was often lost.

Neighbourhood Unit

In the 1930s, Soviet and CIAM urbanism began to diverge with the rise of Stalinism and the decision in 1931 to build the Palace of the Soviets in a neoclassical style, thereby rejecting the modernist approach. Sigfried Giedion, CIAM's general secretary, argued that the pursuit of "order" in urban planning and architecture was a technical question that was "outside of politics" (quoted in Mumford, 2000, p. 88). CIAM's task was to create a "suprapolitical urbanistic 'order'" (Mumford, 2000, p. 91). The divergence of CIAM and Soviet ideas was perhaps marked symbolically by a "protest collage" that a frustrated Giedion made and sent to Stalin himself to express CIAM's disapproval of his decision – CIAM members in the USSR claimed that it was only out of sheer luck that Stalin never got it in the end (Mumford, 2000, p. 73).

However, the 1935 General Plan for the Reconstruction of Moscow still reflected CIAM influences (Mumford, 2009, p. 246). The *kvartal*, often referred to in English as a superblock or neighbourhood unit, drew both on May's CIAM functionalism, where these settlements were planned as mini-communities, as well as Stalinist classicism, in that the *kvartaly* were usually closed blocks, forming inner courtyards, similar to the Red Vienna projects, rather than the long rows of apartments characteristic of CIAM's rational approach to building. It was not until the 1940s and early '50s, argues historian Stephen Kotkin, that the Soviet kvartaly came into their own in what he calls the "golden age" of Stalinist architecture (1996, p. 244). According to Czech architect Jiří Kroha, a kvartal would house anywhere between 4,000 and 6,000 people ([1935] 1969, p. 127).

Eric Mumford contrasts the Soviet kvartaly with the "American norm": "small single-family house subdivisions located in single-use zones well away from city centres" (2009, p. 246). Although Mumford does not mention him, Clarence Perry and his 1929 urban plan for what he called a neighbourhood unit was a key design principle in the North American equivalent of the kvartal. There, the neighbourhood unit was an architectural and urbanistic response to the spread of the automobile and it rejected the grid streets of the past in favour of an interior street system that minimized through traffic. It was as influential as Wright and Stein's Radburn plan – resonating in Europe, including in London and Stockholm (Ward, 1999), and in Canada, and Toronto specifically (White, 2016). Peter Hall ([1976] 2002) notes that the influence of the neighbourhood unit in post-war Britain, especially in the new towns, "is everywhere to be seen" (p. 38).

Perry argued that the actual design of a neighbourhood unit was "forced by the automobile" and that "arterial highways must necessarily run in every direction and turn the street system into a network" (1929, pp. 30, 31). Housing would occupy the "interstitial spaces" in neighbourhood "cells": "the cellular city is the inevitable product of the automobile age" (Perry, 1929, p. 31). The "automobile menace" was a "blessing in disguise" because it called attention to the need to standardize this neighbourhood cell or unit, which would be bound, but not penetrated by, the street network. Residences would make up the interior of the unit, although shops within walking distance would be placed at the intersections of streets on the edge of the unit, an element that Peter Hall notes was not taken up by British post-war planners ([1976] 2002, p. 38).

There was also a moral component to the neighbourhood unit in the North American context in the way that it envisioned collective life: groups of single-family houses surrounding a school and neighbourhood institutions. In Perry's iconic diagram of the neighbourhood unit, the shops and any apartments would be relegated to the periphery of the neighbourhood at the intersections of the "main highway" and "arterial street." In his proposals for "low-cost suburban development," Perry envisioned 160 acres for a neighbourhood unit, with 822 family houses compared to 147 apartment suites. Echoing the separations that in part defined modernist architecture and urbanism, planner Paul Hess (2005) writes that apartments were used as a "buffer" between the single-family homes on one side, and retail and arterial roads on the other (p. 31). Perry did offer a neighbourhood unit solely devoted to apartments, with an estimated population of 10,000, but these were intended for downtown "slum" areas in need of "rehabilitation," not the periphery.

Urban historians (e.g, Wakeman, 2016) have also argued that the neighbourhood unit concept was behind many of the early post-war socialist housing developments. The *mikroraion*, the socialist equivalent of the neighbourhood unit and the successor to the kvartal, dominated architectural and urban planning in the Communist Bloc well into the 1970s and '80s. In her work on Romanian socialist architecture of the 1950s and '60s, Juliana Maxim (2009) notes that "the *mikroraion* shared many of its principles (and origins) with the western notion of the neighbourhood unit" (p. 10). Although the automobile was not the "menace" it was in North America, Maxim notes that like in North American planning, building in the mikroraion "no longer stands in relation to the street, but to the neighbourhood" and the street is no longer the main focus of urban life (p. 12). Writing on socialist Sofia, Hirt (2017) similarly describes the mikroraion as a "multi-functional, if not fully self-sustainable neighbourhood unit" (p. 72) inspired by the Athens Charter (p. 73).

However, Maxim contends that calling socialist developments a variation of the neighbourhood unit robs them of their specific socialist content: the socialist city-building project planned by a centralized state, and the very architecture of the mikroraion – the uniform apartment blocks – are not, Maxim argues, simply a result of new industrialization technologies, but rather an effort to achieve a "'collectivization' of buildings" that mirrored the collective goals of socialism, rather than individualist-oriented ones of the Radburn suburb – that is, the mikroraion is inseparable from the larger *raion*, the region. If the neighbourhood unit and the superblock in North America was about separating pedestrians and cars but also connecting suburbs to the surrounding city through car trips, and affirming the single-family house as the ideal space for a family, the kvartal and the mikroraion were not just suburbs of the city, but essential elements in the wider building of socialism (Zarecor, 2018, p. 9).

Post-war CIAM and the Heart of the City

The turn towards socialist realism in the newly formed Eastern Bloc countries after the Second World War further increased the divide between CIAM and socialism. In the relationship between East and West, and between CIAM's "apolitical modernists" and the socialist architects (both those working within the Eastern Bloc and the Soviet Union), the tension takes on a new form, beginning with the seventh CIAM congress in Bergamo, Italy, in July 1949. The key discussion, from which Sigfried Giedion offers an excerpt in *Architecture, You and Me* (1958), was in part about the importance of city centres. In the discussion, CIAM president Josep Lluís Sert argued that the cities of the world lack civic centres where people can "walk freely and look around" free of car traffic and advertising. The crux of the conflict came with the then vice-president of CIAM, Helena Syrkus (Poland), who came out in support of socialist realism, rejecting modernist architecture as the style suitable for the rebuilding of a bombed-out Warsaw. "We of CIAM must revise our attitude," she proclaimed, and the Eastern Bloc countries should "have a greater respect for the spirt of the past" (Giedion, 1958, p. 87). This approach, rather than modernism and its rejection of the past, would guide the post-war rebuilding of Warsaw. How did Giedion react? Modernist architects also have a "love of the past," he said, "but it is quite another matter to hang a façade of colors or a fresco over the front of a building. This kind of things is for us pure reaction" (1958, p. 89).

Czech architectural historian Marcela Hanáčková writes that in the wake of this split, there emerged two approaches to the question of the city centre. One would come to be the focus of the CIAM 8 in Hoddesdon, in the United Kingdom, in 1951. There, the focus was the "Heart of the City" and the modernist rebuilding of city centres. The second approach, socialist realism, was historicizing and focused on national values. The reconstruction of bombed-out city centres following socialist realism would recreate the city as a copy of the past without the help of modern architecture (Hanáčková, 2011, p. 203). The split between the Eastern Bloc countries – specifically, Poland, Hungary, and Czechoslovakia – and CIAM over socialist realism meant that no Czechs would participate in the CIAM congresses between 1950 and 1955; they returned for the 1956 congress in Dubrovnik, Yugoslavia (Hanáčková, 2014, p. 74). In the proceedings for CIAM 8, Czechoslovakia, along with Hungary, Poland, and Yugoslavia, are listed as "Groups under Reorganisation."

CIAM historian Eric Mumford notes that CIAM 8 was the most important of the post-war CIAM congresses and as the balance of this chapter suggests its ideas travelled and resonated in North America as well as the Eastern Bloc countries, even if these countries were not present at the actual congress. It attempted to reconcile the "twin and potentially conflicting notions of 'urbanity' and 'mobility'" (Gold, 2006, p. 113), and in particular the rights of the pedestrian against the rights of the automobile. In the conference proceedings, Giedion (1952a) argues that the rights of the pedestrian to be at "the centres of community life" have now been "overridden by the petrol engine and so the gathering places of the people ... have been destroyed" (p. 18). Giedion (1952b) insists that vehicle traffic be kept out of the core: the "reconquering of the right of the pedestrian ... is the first requisite of the contemporary city plan" (p. 161). Giedion had already offered his take on the larger problems of the mechanization of everyday life in *Mechanization Takes Command* ([1948] 1969) and on the need to "return to the human scale" and assert "the rights of the individual over the tyranny of mechanical tools" (p. 127). *Mechanization Takes Command* studies "anonymous history" – the unknown objects and inventors of modern industrial society that "have shaken our mode of living to its very roots" (p. 3). Giedion describes these objects as "modest things of daily life, they accumulate into forces acting upon whoever moves within the orbit of civilisation" ([1948] 1969, p. 3). As a response to the title of his book, Giedion calls for a balance between the domains "fit for mechanization and those that are not" ([1948] 1969, p. 720).

Beginning with CIAM 8, the Athens Charter dogma of separating functions came under increasing scrutiny, particularly the over-privileging of vehicle circulation to the exclusion of public and community spaces. The British urban planner Jaqueline Tyrwhitt, one of the co-organizers of the congress, critiqued the Athens Charter's emphasis on "bands of separation" in favour of "centres of integration" (1952, p. 104). In his essay on this congress, Welter (2003) argues that it signalled a movement away from the strict functionalism and rationality that CIAM had propagated in the 1920s and early '30s. This critique of CIAM's rational and functional principles had been building in the post-war period. Sweden's "humanized" functionalism, for example, eschewed the "modern monumentality and the 'heroic' use of materials" typical of CIAM's approach to architecture, and instead emphasized the picturesque, different materials, and a mix of low- and high-rise buildings (Mumford, 2000, p. 167).

Architectural theorist Judith Hopfengärtner (2017) suggests that it offered "alternative strategies of humanization within the capitalist system" (p. 15) – the subtitle of the congress was "towards the humanization of urban life." In a later essay on "The Humanization of Urban Life," Sigfried Giedion (1958) wrote that the idea of the core came from the British CIAM group – Modern Architectural Research (MARS) – in place of the more bland "civic centre," which was too focused on administrative spaces rather than on the elements that "makes a community a community and not merely an aggregate of individuals" (p. 127). The core was more than just an actual planned space in the city, like a pedestrian plaza or a civic centre, argued Tyrwhitt, but rather any space in the city that might serve as a gathering place (1952, p. 103). Urban planning should create spaces that will encourage what CIAM saw as the main aspect of the core: a "rendezvous" (Tyrwhitt, 1952, p. 165). In his contribution to the CIAM 8 publication, Dutch architect J.B. Bakema writes that the "moment of the core" can exist in any place – even a Finnish sauna – at any moment of interconnection when people "discover the wonder of relationship between man and things" (1952, p. 67). This was an explicit critique of the idea that recreation – one of the Athens Charter's four functions – had a fixed space in the city. Tyrwhitt and the younger architects of the MARS group, consisting of mostly British architects who would later form Team 10, considered automobile-dominated, low-density urbanization a result of the "functional differentiation of land use and the alluring imagery of the 'city in the park'" (Gold, 2006, p. 113).

Although one of the impetuses of the Heart of the City theme was the need to rebuild bombed-out city centres – the Lijnbaan shopping street in Rotterdam designed by Team 10 architects Johannes Van der Broek and Bakema is a frequently cited example (see figure 2.7) – many of the designs presented at CIAM 8 looked at new cities in places not affected by the war. In its focus on the "recentralization" of urban space, the meeting attempted more generally to curb "suburbanism" and "unplanned decentralisation" (Sert, 1952, p. 4). Here, Vällingby is again a key example, this time with its town centre (shown in figure 2.8). The socialist-realist approach also did not concern itself with simply rebuilding historic city centres, even if this was the source of the bifurcation within CIAM. As the next chapter will show, the early

Figure 2.7. Lijnbaan pedestrian shopping street in Rotterdam. Source: John Reps Papers, #15–2-1101. Division of Rare and Manuscript Collections, Cornell University Library

socialist suburb of Poruba was an example that applied the historicizing tendencies of socialist realism in building an entirely new district.

Focusing on the core of the city, architects and planners sought to reconcile the scale of the pedestrian with the scale of the automobile – namely, by continuing to provide for the necessary infrastructure of the automobile while also offering an alternative space to the street as a gathering place. More generally, post-war modernist urbanism sought to combat the all-out attention to automobile circulation, not by questioning the dominance of the automobile, but by creating car-free spaces either above ground or, more appealingly, at street level. CIAM's point was not to question separation outright – some separation is and continues to be necessary as long as there continues to be highways, and even the harshest critics of CIAM can be the strongest supporters of physically separated bike lanes. The point is to look at the ways in which that separation was imagined, and clearly there was a diverse range of responses – as diverse as modernism itself.

Figure 2.8. View from the Vällingby town centre. Source: John Reps Papers, #15–2-1101. Division of Rare and Manuscript Collections, Cornell University Library

Modernism Thinks Big

On 7 December 1954 Soviet president Nikita Khrushchev gave a speech to the All-Union Conference of Builders, Architects, and Building Industry Workers in which he extolled the virtues of the industrialization of building that many of the architects of the 1920s and '30s supported. Khrushchev claimed that "given concrete, electric motors, and lifting cranes, and other machinery – it is impossible to continue to work in ancient ways" ([1954] 1963, p. 161). Khrushchev called for standardized designs in building, and he critiqued those architects who would rather "build monuments to themselves" ([1954] 1963, p. 165). He also criticized socialist realism, with its "needless adorning of facades" and "unnecessary decorations" and architects who claimed that they were rejecting constructivism and the "dull 'box style' characteristic of modern bourgeois architecture," but reacted by "decorat[ing] building facades excessively ... thus wasting state funds" ([1954] 1963, pp. 168, 171).

No apartment building should be a "replica of a church" – the architect might need "beautiful silhouettes," argued Khrushchev, but "the people" need apartments ([1954] 1963, p. 170).

Khrushchev's speech also encouraged architects and planners to learn from and indeed one-up the West, and so the "tirades against modernist architecture and urbanism of the late 1940s and early 1950s" that kept the Eastern Bloc countries from the CIAM congresses gave way to a "business-like interest" in the urbanism of the West (Beyer, 2011, p. 77). In one sense, this merely reaffirmed a practice that was ongoing and could be traced back to Soviet urbanists' 1930 declaration in the pages of *Sovremennaya Arkhitektura* (*Contemporary Architecture*) to "learn … not just to copy capitalist achievements, but to discover in the most advanced capitalist technology the seeds of a new socialist technology" (quoted in Kopp, 1970, p. 250). In *Town and Revolution* (1970), Anatole Kopp argues that there is a social content to modernism worth rescuing, one that was outside of and offered a critique of capitalist modernization and, as such, capable of producing a new kind of space (Stanek, 2011, p. 149).

In the wake of Khrushchev's speech and his rejection of socialist realism, the countries of the Socialist Bloc returned to CIAM in time for the 1956 congress in Dubrovnik. Although the Czechs did not participate in CIAM 8, the ideas, particularly those proposed by Jaqueline Tyrwhitt and the MARS group, would form a beginning point for Team 10, the breakaway group of architects attempting to reinvigorate modern architecture with new ideas. Hopfengärtner (2017) writes that the visions for the core of the city in Central and Eastern Europe bifurcated between the "heroic gesture" of Sigfried Giedion's visions for monumental civic centres and the Team 10 participants, who sought an alternative to the "almost exclusively rationalist orientation of modern architecture." Although Tyrwhitt and others had already offered a critique of this rationalist approach, it was still on offer in the CIAM 8 proceedings. The collectively authored piece that appeared in the conclusion to the conference proceedings hardly evokes the spontaneity associated with bustling downtowns: "the authority should endeavour to put sufficient material means at the disposal of the citizens to enable them to display and declare their spontaneous reactions" ("Attributes of the Core," 1952, p. 166).

Lukasz Stanek and Dirk van den Heuvel (2014) suggest that the architects from Central and Eastern Europe shared their Western colleagues' critique of Athens Charter urbanism, which they felt did not speak to the post-war situation: "technological progress, personal mobility, the increasing importance of leisure, varying scales of human associations and multiple modes of belonging" (p. 12). These elements reverberated in both post-war Canada and Czechoslovakia, and specifically in the plans for South City and Willowdale, and the attempts to offer an alternative to Athens Charter urbanism and to bring density, centrality, and collectivity to the urban periphery. Czech CIAM historian Marcela Hanáčková has noted the connection between Team 10 and the work of South City's chief architect, Jiří Lasovský. His earlier design for a pedestrian shopping street in the Pankrac neighbourhood of Prague (shown in figure 2.9) was influenced by his visit to the Lijnbaan city centre.

Hopfengärtner (2017) notes that Team 10's humanism often clashed with the increased standard of living in both East and West, which entailed new-found leisure time and the increased consumption of consumer products. This was certainly the case with the automobile, which saw large increases in ownership in the 1960s and '70s even in socialist countries like East Germany and Czechoslovakia. Reconciling the growth in automobility with the need for pedestrian spaces began in earnest with CIAM 8, but it continued in the 1950s and '60s in the West and in the Eastern Bloc countries, with plans for city centres of various kinds as well as the utopian visions of urban megastructures that offer a vision of collective public spaces.

Although Team 10 largely grew out of the European context, the importance of the core continued in North America, as Josep Lluís Sert, the CIAM president at the time of the Heart of the City congress, would go on to head the Graduate School of Design at Harvard from 1953 to 1969, where he "continued to push the agenda of CIAM 8" (Yoos & James, 2016b). Eric Mumford (2009) argues that Sert brought together the ideas of the old-guard modernists with the new ideas coming out of the Team 10 group. In 1964, Sert and Tyrwhitt organized the Harvard Graduate School of Design eighth-annual urban design conference on "The Role of Government in the Form and Animation of the Core."

Charting the circulation of ideas behind the "multilevel metropolis," the architects Jennifer Yoos and Vincent James point to the key role played by the first Harvard Graduate School of Design conference in 1956. One of the participants in this conference was Austrian émigré architect Victor Gruen, who presented his plan for redeveloping the downtown of Forth Worth, Texas. Gruen made the "heart of the city" the main subject of his 1964 book, whose title – *The Heart of Our Cities* – is a seeming reference to CIAM's work (the CIAM 8 proceedings are listed among "Some Books of Special Interest"). Gruen, best known as the designer of the first suburban shopping mall, was a strong advocate of separating cars and pedestrians in rebuilt city centres and building regional suburban shopping centres as alternative

Figure 2.9. The Pankrac district pedestrian area, designed by Jiří Lasovský. Source: *Architektura ČSR, 40*(6) (1981)

public spaces (surrounded by vast parking lots). Gruen's two major interventions – the regional shopping centre and the downtown pedestrian mall – focused on public spaces for gathering, not just spending money. The public, open spaces of the shopping centre were to be the site of cultural activities and political gatherings (and his Midtown Plaza in Rochester was cited as one of the influences for the initial plan to redevelop Willowdale). Gruen's plans for Fort Worth won the accolades of activist and critic Jane Jacobs, also attending the 1956 conference (Yoos & James, 2016a, p. 98). Jacobs is best known for *The Death and Life of Great American Cities* (1961), in which she lambasts the top-down planning of Ebenezer Howard, Le Corbusier, and Frank Lloyd Wright, emphasizing instead what people are doing down on the street and on the sidewalk. In an earlier essay, she set out some of these ideas in assessing projects around the United States to revitalize downtowns. Her assessment? "They banish the street. They banish its function. They banish its variety" (Jacobs 1958). The one exception, she argues, is Victor Gruen's redevelopment plan, not for its attempts to deal with traffic, but for the way his plans for a car-free section of downtown enlivened rather than killed the existing streets. She praised the plan for the way it focused on "compact and concentrated" activities around the existing buildings, adding pedestrian arcades, outdoor cafes, kiosks, and "special lighting effects." This was also a focus of CIAM 8, and in particular the importance of activity spilling into squares and pedestrian areas. Strictly opposed to CIAM's separation of functions, Jacobs appreciated the way Gruen worked with activities at the street level, instead of putting all life either above or below ground (even though the plan did include elevated pedestrian walkways that would connect the centre to the parking garages, situated on the periphery of the pedestrian zone).

Gruen's Forth Worth plans also appeared in Soviet and Czech publications at the time. In the Soviet context, writes architectural theorist and historian Elke Beyer, Gruen's plan was primarily taken up as a technical question: how best to organize car traffic and parking lots (the aspects Jacobs was least interested in). Beyer writes that the socialist city centres as "spaces of communication" were often defined by their technical approach to traffic and pedestrian despite some of the fantastic visions of the city offered by Soviet avant-garde architects. Following Khrushchev's speech in 1954, Beyer writes, planners and architects "claimed authority as technical experts working with scientific methodology rather than as artists," focusing on transport and traffic planning (2011, p. 89). Traffic was important, and the pedestrian-only areas were seen as key aspects of the socialist approach to the city centre, but the city centres in the socialist context also sought an alternative to the individualist consumerism of the West, and architects did think of themselves as artists sculpting an environment. In *Budoucnost měst* (*The Future of Cities*), Jiří Hrůza also cites Gruen's Forth Worth plan as well as his suburban "shopping centres." There was a clear distinction, at least in Hrůza's mind, between the American shopping malls and the Swedish and Dutch examples of pedestrianized shopping streets (although the first shopping malls that appeared in Sweden and Finland were in Vällingby and the new town of Tapiola). For Hrůza, the American city centres were focused on offering the best choice of goods in an area with plenty of parking, but socialist centres could not and would not be simply about "solving transport problems and freeing up more space for shopping malls and administrative buildings" (1962, p. 187). Discussing a Soviet work of urban planning about city centres, Beyer (2011) remarks how Soviet planners could openly talk about American plans for transport engineering, but at the same time not mention that these plans were so clearly rooted in individualism, consumption, and capitalism. Hrůza, on the other hand, explicitly calls out this commercialism and rampant automobility as a way to distinguish the socialist city centre, even if architecturally or technically they might resemble their American counterparts. The social and communal (*společenský*) centres were not just "analogs of American shopping centres," writes Hrůza, but "centres of social life" for the city's residents (1962, p. 189). The city centres should reflect the principles of "socialist humanism": improve housing conditions, offer access to green spaces, and create the socialist city as work of art with murals, water parks, and sculptures, but without reproducing the decorative tendencies of socialist realism. It should be easy to get to, but with the "most important spaces reserved only for pedestrians." There were also undertones of social engineering as Hrůza writes that it will be necessary to "eliminate all the hygienic flaws and remove everything that does not belong in a centre" (1962, p. 187).

If the city was to be a work of art as much as it would facilitate the movement of cars, pedestrians, and public transit, the utopianism of the 1960s, embodied in the megastructure and the "multilevel metropolis" (Yoos & James 2016b) brought this to a new level, both literally and symbolically. The megastructure in particular was the answer to CIAM's call at the 1951 congress to create a monumental architecture and an architecture of "urban 'spontaneity'" that would also emphasize fun, flexibility, and transience. The main motivation behind megastructures was to contain the exploding metropolis and the increase in automobile ownership by concentrating as many activities as possible under one roof (Banham, 1976, p. 199), including shopping, residential, and civic functions. In *Megastructure: Urban Futures of the Recent Past* (1976) architectural critic Reyner Banham calls this approach "a defiant gesture in favour of an older type of urbanism" (p. 170) and also a critique of Athens Charter urbanism, even if the results were not going to look anything like that older urbanism.

This utopian gesture was mirrored in the socialist countries. "The twenties are being taken out and dusted off," writes Anatole Kopp (1970), to serve as "a point of departure for the research now in progress" (p. 236). A 1967 feature on 50 years of Soviet architecture in *Architektura CSSR* drew connections between the work of the 1920s and early '30s and the Soviet visions of the urban region in the 1960s and in the future. An article by historian and architectural theorist Andrei Ikonnikov featured dynamic drawings of kinetic cities similar to those of Archigram in the United Kingdom.

The megastructure as the built form of these utopias was the most monumental of all these post-war visions and its most well-known iteration was the town centre of Cumbernauld, a new town situated between Glasgow and Edinburgh for around 50,000 inhabitants on almost 1,700 hectares (Forsyth & Crewe, 2009, p. 425). Although Cumbernauld was a new town, and not necessarily a suburb, the architects of both South City and Willowdale cited it as an influence on their designs of their respective suburban centres. These ideas circulated through and informed South City's central area and the core features of Willowdale's redevelopment. Cumbernauld's town centre megastructure is the suburb's most noted, and most loved and loathed, contribution to modernism (see figure 2.10). From the beginnings of its planning in the late 1950s, the town centre was asked to accommodate both pedestrians – it should be accessible on foot to all residents of Cumbernauld – and cars, as it would occupy a "nodal position on the road network to allow speedy and efficient movement of people and goods to and from the centre" (Gold, 2006, p. 150). In his essay on Cumbernauld, the geographer John Gold suggests that the megastructure would tempt suburban dwellers back to the city by offering new, modern environments, although the megastructure was not confined to city centres, but brought out into the expanding urban region, to offer density amidst sprawl.

One of the main elements of most megastructural schemes was the paradoxical "burial" of the automobile problem in high-rise parking lots. The point was to "dispose of the automobile" (Banham, 1976, p. 43), not by banning it from the centre, but by hiding it in elaborately designed above-ground parking lots, such as the twin towers in Marina City, Chicago (1962), rather than leaving it open and visible in surface parking lots (Banham, 1976, pp. 170, 40). At the same time, Cumbernauld's town centre, which straddles the motorway, did anything but hide its connection to cars: it gave drivers the "Futurist experience of plunging through a vast urban structure" (Banham, 1976, p. 170). What makes modernist urbanism more complex and so often contradictory is that its architects and urban planners were in part reacting against automobility while their responses simultaneously assured the spread of automobility. Modern urban planners envisioned the enhanced mobility of both drivers *and* pedestrians, part of which would include extensive pedestrian infrastructure even if that infrastructure was isolated in a sea of cars.[4] The megastructure reflected a wider need for "a more 'intense' urban experience" and it challenged the dominance of Athens Charter urbanism (Banham, 1976, p. 126).

The modernist urbanism of the megastructure may be formulated as the "quotidian made spectacular." The utopian visions of 1960s modernist architecture literally and symbolically elevated the everyday practices of walking, shopping, and hanging around. And it was also its spectacular designs that doomed many modern projects, like Cumbernauld, which was not built according to design. The ambitious nature of so many megastructure projects often guaranteed that they would

Figure 2.10. Pedestrian entrance into the town centre megastructure in Cumbernauld. Source: John Reps Papers, #15–2-1101. Division of Rare and Manuscript Collections, Cornell University Library

remain unbuilt. Still, there were also the ideas behind the designs that explicitly sought to respond to the fractured and fragmented spaces produced by CIAM. Concluding his essay on Cumbernauld, Gold writes that there was a "body of thought within the Modern Movement that challenged the orthodoxies with which that movement is normally associated and that looked to find other, more socially responsive living environments" (2006, p. 126).

Conclusion

In *Militant Modernism* (2008), Owen Hatherley argues that returning to modernism does not mean treating modernism as heritage, the architecture of which should be preserved, but looking to the visions and ideas behind it, so much more than the architectural style of an individual building, to which modernism is so often reduced. The reconsideration of the "left modernisms" of the 1920s and their renewal in the 1960s gives an alternative and more nuanced reading of the modernism of the interwar and post-war periods, encapsulated in the title of Anatole Kopp's 1988 work: *Quand le moderne n'était pas un style mais une cause*. Hatherley finds in the "concrete walkways and windswept precincts" of the 1960s British megastructures of the new town of Cumbernauld in Scotland or the Barbican in London, as Benjamin did in the arcades of the nineteenth century, a nostalgia for a modern future that did not quite happen as it was supposed to. In *A New Kind of Bleak*, Hatherley writes that in the Cumbernauld town centre megastructure we can find "glimpses of potential new worlds" (2012, p. 1). Given that these structures were produced at the height of British post-war social democracy, Hatherley argues that they offer a critique of present-day inequalities, particularly as regards to architecture. The modernism of the 1920s and the 1960s was "immersed in the quotidian," while today architecture that still calls itself modernist is based on the spectacular and a "distance between itself and everyday life" (Hatherley, 2008, pp. 8, 12). The modernist urbanism and architecture described in this book is not only the result of the heroic architect designing singular buildings; it is also the result of the collective projects that sought to transform everyday life.

It is with this sense that we can build on and develop Gold's reference to the "socially responsive living environment" as a jumping-off point for discussion of the alternatives to the car-dominated suburbs, and to the lack of centrality in the suburbs. The problem, however, was reconciling this desire for socially responsive living environments with increased standards of living and the increased prevalence of cars, but also the increasingly technocratic approach to building new towns and revitalizing downtowns. Yoos and James (2016b) explain why Gruen's innovative street-level plans never did come to fruition:

> At the height of suburbanization in a car-crazed country, his plans for urban pedestrian zones were rarely implemented as he envisioned. Instead, city planners and developers spanned streets with footbridges, or tunneled under them, creating a more unified retail experience without threatening the primacy of automobiles.

And yet, in both the interwar years and the post-war socialist context, there were many attempts to offer an alternative vision of the suburb that was modern but without the strict focus on cars and single-family homes. These alternatives are particularly important given that one of the twentieth century's key architectural works – Sigfried Giedion's *Space, Time and Architecture*, which had five editions from 1941 until 1967 – makes no mention of Soviet architecture and urbanism, nor does it mention the Red Vienna housing projects, which saw 4,000 apartment blocks with collective facilities built between 1919 and 1934 by many noted international modern architects (Blau, 1999, p. 8). The Vienna housing projects were very much a part of the lineage of modernist urbanism, even though they did not subscribe to some of its key tenets, like Stein and Wright's Sunnyside Gardens and Radburn, which Blau puts at the source of Giedion's rejection. Vienna eschewed the "Taylorized living environments of the German *Siedlungen*" (Blau, 1999, p. 7), so the latest mass production technologies were not used, and decoration was important. They were often purposefully "woven into the existing fabric of the city" rather than grown out of the sundering of the street, even though in many cases certain streets were closed to car traffic within the superblocks. Red Vienna was antithetical to modernist urbanism even though it clearly grew out of it. As the architect Hubert Gassner, describing Vienna's superblocks, put it: "nicht modern, nicht antikisch."[5]

More pertinently, Giedion did not mention any of the architecture of the Eastern Bloc countries, including Czechoslovakia. Giedion knew this work existed as he worked closely, through CIAM, with the architects and theorists from Czechoslovakia, including Karel Teige, and the history and conflicts within CIAM were guided by the interactions of architects from East and West. The Functional City that was the topic of the first congress, held on the boat to Athens, was originally supposed to take place in Moscow in 1932; similarly, Prague was the original choice for CIAM 7 in 1949, but the venue was shifted to Bergamo, Italy because of East-West tensions (Mumford, 2009, p. 248).[6] It is possible he chose not to include the socialist projects because of the climate of anti-communism in the United States, where he gave the lectures that make up *Space, Time and Architecture*. That this book helped define the cannon of modernist works of architecture and planning (Berman, 1983, p. 302) makes it especially important to understand its partiality, and its exclusions (Spechtenhauser & Weiss, 1999, p. 246). The Czech modernist architect Karel Havlíček thought that Giedion was fomenting a "Cold War" within CIAM by ignoring the architecture of Central and Eastern Europe.[7]

The story, of course, is more complicated than the extent or limits of Giedion's social circle. The interwar period, particularly the 1920s, produced a flourishing of modern architecture and urbanism in Czechoslovakia. The history of modernist urbanism in Czechoslovakia, and its implications for the urban periphery, offers a parallel to the story told here, and it begins with the formation of the state of Czechoslovakia in 1918 and the work of one of the European avant-garde's most influential figures, Karel Teige.

3 Socialist Space

If one of the aims of the previous chapter was to show how socialist and capitalist ideas were inextricably intertwined, beginning with CIAM in the late 1920s and culminating in the utopian discourse of the 1960s, then does it even make sense to ask if there was a socialist space distinct from the capitalist one? One of Henri Lefebvre's central hypotheses in *The Production of Space* is that "each mode of production has its own particular space" and that "the shift from one mode to another must entail the production of a new space" ([1974] 1991, p. 46). Lefebvre further argued that the "technological and technocratic rationality" of the functional zoning that was the hallmark of the Athens Charter was a projection of both state capitalism and state socialism ([1974] 1991, pp. 317, 304). Writing in the shadow of the suppressed uprisings in Prague in 1968, Lefebvre ([1970] 2003) finds that the idea of the urban in socialist countries did not differ "in any significant way" from the urban revolution in capitalist countries (p. 111) and that "urban space is no differently defined in a socialist country than it is elsewhere" (p. 113).

This chapter does not try to give a definitive answer to Lefebvre's question. Rather, building on the work outlined in the previous chapter, it probes the nuances of the differences between capitalist and socialist urbanism by examining four iterations of socialist space in Czechoslovakia: that of the interwar period, the early post-war period, the period between Expo 58 in Brussels and Expo 67 in Montreal, and finally the so-called normalization period of the 1970s and '80s following the Russian occupation of Czechoslovakia in 1968. Each of these iterations offered an alternative to the dominant vision of the suburb – a privatized life lived out in a single-family house – while at the same time reflecting influences from the architecture and urban planning of the West. Each period illustrates the complex way that modernist urbanism was taken up in the socialist context.

With the formation of the socialist states of Eastern Europe in 1948, the socialist city could now be realized outside the Soviet Union. Theorists from various disciplines have long considered the question of the socialist city's autonomy from the West (e.g., French & Hamilton, 1979), but this was taken up in earnest beginning in the 1990s as the features of the socialist city began to disappear and cities began to adopt the neo-liberal tendencies of their counterparts in the West (Szelényi, 1996, p. 288).[1] Did the socialist mode of production produce its own autonomous space distinct from the capitalist mode of production? Or was urbanization in East and West connected to the wider processes of industrialization and modernization? If it was the latter, then the differences between the socialist city and the capitalist city were simply a question of differing levels of industrialization and modernization – that is quantitative differences, rather than some qualitative difference between socialism and capitalism. In this sense, the main determining factor was levels of urbanization and industrialization rather than the mode of production. The sociologist Iván Szelényi suggests that socialist cities *did* exhibit remarkably similar features distinct from the cities outside of the Eastern Bloc. However, these very features may not have been intended by the socialist architects and planners – they were the outcome of "the abolition of private property, of the monopoly of state ownership of the means of production, and of the redistributive, centrally planned character of the economic system" (Szelényi, 1996, p. 287).

György Enyedi (1996), taking issue with Marxist political economy, but also agreeing with Lefebvre, claims there was no concrete evidence that suggested the socialist mode of production could address the contradictions of abstract space in a way that was different from the capitalist mode. Architect and urban planner Sonia Hirt (2012) revisits this debate from the perspective of architecture and urban planning, suggesting that if modernist urbanism is understood, according to David Harvey, as "large-scale, metropolitan-wide, technologically rational and efficient urban plans, backed by absolutely no-frills architecture" (1989, p. 66) then the post-war standardized apartment blocks of the socialist city periphery are its quintessential examples. For Hirt, capitalist and socialist states simply offered variants on a twentieth-century modernity committed to massive developments, technological rationality, and a rejection of old building and planning traditions in favour of the new (2012, p. 62).

Lefebvre occupies an interesting position here because when it came to modern architecture and urbanism he lumped together the socialist and capitalist states. Lefebvre used the concept of "the state mode of production" to describe both the socialist and capitalist states, which, as Łukasz Stanek notes, implies that he "admitted that the concept of mode of production is unhelpful for distinguishing between the two camps of countries" (2011, p. 65). Stanek argues that in the 1960s, Lefebvre similarly characterized socialist and capitalist states as "bureaucratic regimes of controlled consumption, oriented toward economic growth" (2011, p. 64). Lefebvre believed that the socialist and the capitalist states were following similar paths, and so produced similar contradictions. More precisely, it was the privileging of industrial production and extensive growth that hindered the emergence of a qualitatively different socialist space. Cultural theorist Susan Buck-Morss (2000) concurs: "socialism failed in this century because it mimicked capitalism too closely" (p. xv).

In "Suburbia in Three Acts: The East European Story," Hirt and Kovachev (2015) bring this discussion out to the suburbs, arguing that twentieth-century Eastern European suburbanization is "sufficiently unique to warrant telling a distinct East European suburban 'story'" (p. 177). They point to three separate acts in this story: the period before socialism, and especially the interwar period; socialist suburbanization (1948–89); and post-socialist suburbanization. They argue that this is a "unique" story because of the three different governing types of each period. They devote the bulk of their story to the third act, but the complexity of post-war socialist suburbanization in particular demands a fuller treatment even if the end product of that period – row after row of prefabricated apartment blocks – would suggest that the story is a simple one of state-led and state-dictated architecture.

Although the First Czechoslovak Republic was capitalist, it still leaned strongly toward the left. The Soviet urbanism and deurbanism of the 1920s and early '30s had an important influence on socialist city building, particularly the large-scale developments on the periphery of cities. Architectural historian Kimberly Elman Zarecor (2011), whose work on post-war Czechoslovak urbanism is a key reference point for this book, also points to the interwar architectural debates in Czechoslovakia as central to post-war architecture – far from peripheral, the Czechoslovak interwar avant-garde was active and known throughout Europe. Although more focused on the post-socialist environment, geographers Kiril Stanilov and Luděk Sýkora (2014) similarly argue that the "intellectual inspiration" for the post-war socialist suburbs was the urban transformations dreamed up by the architects of the 1920s and '30s (p. 259).

Although the socialist countries of Eastern Europe did share some unique characteristics that differed from their Western counterparts, there is no simple distinction between East and West as the architects and urbanists in these countries took part in the increasingly global dialogue of modernist urbanism. This chapter looks at the notion of the "living environment" (*životní prostředí*) as one of the key elucidations of socialist suburban space in Czechoslovakia. Although the architects of Radburn and the neighbourhood unit sought to form urban space as a whole, under socialism, the living environment was not simply a design principle, it was also an approach thoroughly imbued with the goals of socialism. Recent work on the history of Czech architecture by Ana Miljački (2017) and Maros Krivý (2016, 2017) point to the living environment as central to the understanding of architecture in the post-war period in Czechoslovakia, culminating in the UIA congress on architecture and the environment held in Prague in 1968, but also continuing in the architectural debates of the 1970s and '80s. Many architects who were working in the 1960s point to the importance of the UIA congress (Urlich et al., 2006), the theme of which fit well with the "socialism with a human face" emphasized by Czechoslovakia's prime minister during this period, Alexander Dubček. Through the interdisciplinary concept of the living environment, combined with the aspirations of socialism with a human face – which argued that the transformation of the living environment has to be seen along with the promises of humanistic socialism – architects and urbanists alike sought out a qualitatively new socialist space and criticized the deficiencies of existing suburban environments.

The Interwar Avant-Garde

One of Czechoslovakia's most vocal critics of CIAM was the avant-garde theorist and critic Karel Teige, who in the interwar years was "the emissary and representative of the modern movement in Czechoslovakia" and one of the key figures of the European avant-garde (Spechtenhauser & Weiss, 1999, p. 219).[2] Importantly, as a committed leftist, he also engaged with Soviet architecture and urbanism, particularly during the 1920s and early '30s. In their overview of Teige's relationship with CIAM, Klaus Spechtenhauser and Daniel Weiss write that CIAM was never as unified as it may have appeared in documents like the Athens Charter, which itself produced much debate and difference within CIAM. Ideological differences, particularly between socialist-minded individuals like Teige and those who advocated for the functionalist approach at the 1933 congress on the Functional City, was one of the major points of contention.

Teige inaugurated his key role in the European avant-garde with his visit to Paris in 1922, where he met Le Corbusier for the first time; he would later host the architect on his multiple visits to Prague. Although Teige took issue with many of his ideas, he was also an important translator of Le Corbusier's architectural vision. In 1923, he became the editor of *Stavba* (*Construction*), a monthly journal devoted to modern architecture, purism, and constructivism that took up themes that Le Corbusier and Amédée Ozenfant addressed in their journal *L'esprit Nouveau*. Teige translated many of the articles from that journal for *Stavba* and also wrote about Le Corbusier's ideas, in particular his view that a dwelling should be "mass produced and just as available and cheap as a Fiat model 509, a Ford, or a Citroen" (1925–6, p. 139). In that article, entitled "Machines for Living," Teige writes that industrialization makes possible the utopian promises of the past (1925–6, p. 136). The style of the "new architecture" found in cars, airplanes, cinema, and photography is not the result of "aesthetic manifestos" but the "collective and largely anonymous, disciplined ... work of laborers and technicians" (Teige, [1923] 2000, 309). For Teige, the machine for living was an example of "architecture without architecture," which he first used to describe the austere design of a train station (1933, p. 18) and which was central to his visions of a distinct socialist and modern architecture.

In 1924, Teige, along with the modernist architect Jaromír Krejcar, announced the formation of the Club for New Prague. They were responding to the Club for Old Prague, founded in 1900 to better deal with the planned "modernization" of Prague's Old Town (Švácha, 1995, p. 102). In their manifesto, Teige and Krejcar argued for the acceleration and extension of affordable and cheap means of transport. In their view, the form of the modern city should reflect the "modern organization of work." Above all, they claimed that the building of cities was a "scientific, not an artistic problem" ("Club for New Prague," 1925, p. 13).

Teige and Krejcar's New Prague involved a proposal to knock down most of the city's medieval core, save for a few choice historical monuments, and erecting an administrative centre in its place whose streets, unlike the narrow, winding streets of the medieval core, would be able to accommodate modern forms of traffic (Cohen, 2000, p. 38). Teige claimed that "most contemporary cities are useless for modern life" ([1930] 2000, p. 264). He praised the "slum clearance" in Prague, particularly that of the Jewish ghetto, whose winding streets were cleared to make way for a broad boulevard (named, significantly, Paris Street, a seeming homage to Haussmann's campaign in the previous century). With respect to this "necessary revitalization," Teige writes: "this single great urban scheme of century's end was so virulently attacked by members of the Club for Old Prague that the city's bold and often merciless urban development from a medieval town into a modern metropolis was considerably hampered" ([1930] 2000, p. 83). Using the language of surgery typical of heroic modernism (with the architect cast as surgeon), Teige claimed that the city was "sick in all its parts" and that "ingenious traffic regulations" are "local surgeries" that at best "help slow down the disintegrative cancer afflicting our cities" ([1932] 2002, pp. 152–3).

Teige's travels also took him east. In 1925, Teige went to Moscow and Leningrad to see first-hand the fruits of the Bolshevik revolution, meeting with Russian constructivists like Vladimir Tatlin and Kazmir Malevich (Honzík, 1963, p. 72). He was greatly influenced by the Russian constructivist approach to architecture, formulated in the pages of *Sovremennaya Arkhitektura*, which sought "to construct a fitting framework for a developing society which, at the same time, in a dialectical process, would assist that society in its development" (Kopp, 1970, p. 95). Modernist architecture meant nothing if it did not lead to a new way of life. The Russian architect Moses Ginzburg's idea of a "social condenser" was mirrored in Teige's belief that architecture was inseparable from social needs – that is, architecture should be both a "reflection of society" and a "tool for social transformation" (Kopp, 1970, p. 96). These social condensers included workers' clubs, kindergartens, apartments, cinemas, and factories. Blending modernism and socialism, Teige's own take on constructivism "call[ed] for an architecture without cliches, without 'facades' – an architecture that in the course of the construction of socialism will fulfill the role of reshaping our life in all its relationships, an architecture that will provide the blueprint for a new way of life" ([1932] 2002, p. 25).

Teige's belief in modern architecture as a social condenser was behind his scathing critiques of Le Corbusier's plans for the Mundaneum project in Geneva – a "centre for world thought" – which he published in *Stavba* in 1929 and to which Le Corbusier responded later that year. The debate centres around what has become Teige's most infamous claim – namely, that architecture should create instruments, not monuments ([1929] 1974, p. 90). To call itself modern, argued Teige, architecture should be dictated by actual social needs, not by monumentality, which only leads to the "monstrosities" of palaces and castles ([1929] 1974, p. 89). An "affection for art" had hindered the architect's ability to build workers' housing, apartments, and schools (Teige, [1929] 1974, p. 91). In this vein, Teige levelled one of his characteristically brazen criticisms of Le Corbusier, claiming that his Swiss colleague "sins against harmony, having formulated such a clear and comprehensible notion as the 'machine for living,' he depreciates it by adding vague attributions of dignity, harmony and architectonic potential, through which he can then embrace all aestheticism and academicism" ([1929] 1974, p. 89). In "Modern Architecture in Czechoslovakia," Teige writes that modern architects should give up their "artistic and individualistic caprice" and "adjust to the conditions of machine production" ([1930] 2000, p. 287). For Teige, architecture ceases to be architecture in the historic and aesthetic sense of the word, and instead becomes *stavba* (construction) and *stavební tvorba* (building production) ([1930] 1977, p. 171).

The debate with Le Corbusier highlights Teige's commitment to the idea of a machine for living and an architecture that creates instruments, an architecture that partakes of the ideas coming from both the East and the West. Whereas Le Corbusier seemed to share Teige's concern that social needs dictate what gets built, he also sought out an architecture that through the pursuit of order, composition, and harmony would stand outside of politics – hence his willingness to work for regimes all over the political spectrum (e.g., Weber, 2008): for Le Corbusier it was architecture *or* revolution (and he chose the former). In an essay on the sociology of housing, Teige rejected this position, arguing that "the housing question is unsolvable without a revolution" and the abolition of private property ([1930] 1977, p. 166). Teige, along with those of his colleagues who attended the 1929 and 1930 CIAM congresses, was committed to a politics in which architecture is the means to socialist ends: architecture alone could not change behaviour – architecture-as-instrument was useless without simultaneous economic and social reform (Baird, 1974, p. 81; Spechtenhauser & Weiss, 1999, p. 233).

The "superficial consensus" within CIAM that the Athens Charter exemplified masked many internal contradictions, particularly the politicized figures within CIAM and those, like Giedion, who advocated a more apolitical, technocratic approach. This difference was at the centre of CIAM, going back to the origins of its very name. In Germany, CIAM was known as the Internationale Kongresse für Neues Bauen, which suggested a "repudiation of *Architektur* ... the art of fine building" in favour of "completely rational building techniques" (Aud van der Woud quoted in Gold, 1997, p. 60). The Czech participants in CIAM preferred Ernst May and Hannes Meyer's approach to "practicality" in building to Le Corbusier's more "conceptual approach" (Gold, 1997, p. 60). The socialist approach was reflected in CIAM's founding document in the section on "Urbanism," which described it as "the organization of all the functions of collective life; it extends over both urban agglomerations and over the countryside." Urbanism should reject any "pretensions of a pre-existent estheticism" and should be about a "functional order" (quoted in Mumford, 2000, p. 25).

In 1933, during the process of editing what would become the Athens Charter, Giedion wrote to Le Corbusier on the fundamental role of CIAM: Would architects be technicians or politicians? Would they "resolve problems on a technical basis" or would CIAM be comprised of "politicians, like the 'Leva Fronta' in Prague," in which "one must profess oneself clearly against capitalism" (quoted in Mumford, 2000, p. 87)? In the letter, Giedion noted that Teige had sent him information on Levá Fronta's work, but that he believed Teige's approach was too simple and it would give CIAM limited influence outside the socialist world.

Levá Fronta was a collective of leftist intellectuals, artists, and architects committed to socialist revolution; the group also had a special architectural section whose members, committed to Teige's way of thinking, would "occupy themselves primarily with the problems of housing ... and the sociological aspects of architecture."[3] The sociological aspects meant that socialist architects critiqued housing above all as an expression of the inequalities of the capitalist system, which could only be addressed by overturning that system rather than simply reforming it, as the mainstream CIAM architects aligned with Giedion claimed (Spechtenhauser & Weiss, 1999, p. 224). The group's composition and aims also signalled an important theme within socialist architecture: interdisciplinary cooperation with the aims of transforming the living environment.

Hannes Meyer, who took over the Bauhaus from Walter Gropius in 1928, invited Teige to give a series of lectures at the Bauhaus on typography and contemporary literature and on the "Sociology of the City and Housing" in 1930 (Spechtenhauser & Weiss, 1999, pp. 235, 251).[4] Teige was a strong supporter of Meyer who brought a new vision to the Bauhaus school focused on the sociology of building. Teige believed that the Bauhaus had become a victim of its international success, which gave rise to the *Bauhausstil*, a "modernisitic fashion" that spread over Europe, "a caricature of the best intentions of the institutions" ([1929–30] 2000, p. 319).

The Garden City

One of the most politicized aspects of modern architecture, in Teige's eyes, was the modern architects who claimed to be interested in social causes and yet still devoted their energies to building villas for the rich rather than apartment buildings for ordinary people. Although multi-family dwellings figured in Stein and Wright's plan for both Sunnyside and Radburn, the focus is clearly on the single-family houses and the parks and gardens around them. In *The Minimum Dwelling*

([1932] 2002), a book that grew out of Teige's contributions to CIAM, he rejected outright the single-family house (and the institution of marriage and family that goes with it) as a solution to housing problems: the single-family house, he argued, is "inconceivable and unjustifiable" as a solution for the housing shortages (p. 102). *The Minimum Dwelling* is Teige's contribution and response to CIAM's second congress on the *Existenzminimum* – low-cost housing for the working classes.

Teige was also reacting against the garden city ideal, which he believed had romanticized the single-family house. The garden city ideas of Ebenezer Howard and Raymond Unwin were a key influence on 1920s Prague urbanism, particularly on the periphery of the city at the end of tramlines or outside of the city along the railroad lines (Sýkora & Mulíček, 2014, p. 134). A full Czech translation of *Garden Cities of To-Morrow* appeared in 1924, one of the earliest translations of Howard's book. In 1921, František Fabinger, the secretary of the Society for the Establishment of Garden Cities in the Republic of Czechoslovakia, privately published *Bytová otázka a zahradní město dle vzoru anglického* (*The Housing Question and the Garden City According to the English Model*) (Guzik, 2009, p. 384).

Prague's garden suburbs, built in the 1920s and '30s, form one of the zones in the city's concentric growth in the twentieth century and were the first attempts at building suburbs (Sýkora & Mulíček, 2014, p. 135). Although not a suburb, the company town built around the Baťa shoe factory in Zlín in the 1920s was fashioned on the principles of functional urbanism and featured a neighbourhood of single-family brick houses owned by the Baťa shoe company and used by workers and their families (see figure 3.1).[5]

Teige's writings, however, were part of a shift in Depression-era Prague urban policy away from building garden cities with single-family houses towards complexes with small apartments affordable to the working classes (Švácha, 1995, p. 304), which would radically reduce private space in the interests of maximizing collective spaces. In the context of his rejection of the single-family house as the solution to the housing question, Teige critiqued the garden city approach to urbanism. Although the "hygienic reforms" of the garden city – health, lower densities, fresh air, etc. – were a response to the overcrowded and unhygienic apartments in the old city, they were not affordable to the majority of those very apartment dwellers (Teige, [1932] 2002, p. 129). Teige wrote that Ořechovka, the most well-known of Prague's garden city suburbs (see figure 3.2), was simply a way for the "Prague bourgeoisie" to elevate their status ([1930] 1977, p. 188). This view was later echoed by Jiří Hrůza (1962), who argued that small garden city developments

Figure 3.1. Family houses built for Baťa factory workers in Zlín. Source: Steven Logan

like Ořechovka did not help improve the overall situation of the city nor did they improve the workers' – for whom Ořechovka was originally intended – standard of living (p. 78). This garden suburb also features a villa by Adolf Loos: the Loos Mueller Villa, now a museum.

In *The Mininum Dwelling*, Teige wrote that the garden suburb is "a false dream, a romantic fallacy, and a dangerous utopia" ([1932] 2002, p. 138). Teige had no shortage of invective for the principle occupants of the single-family house, and for the institution of marriage generally. In his characteristic style (and for which his translator should be commended) he called the marital bed "a hatching place of the most wretched forms of bourgeois sexual life" and "a roosting place of shocking erotic banality" ([1932] 2002, p. 173). In minimum dwellings, on the other hand, each person would be free to form relationships, but always with their own bed to return to at night. (As is common in socialist utopias, children would be housed in separate quarters, contributing even further to the break-up of the single-family household.)

Teige singles out the family house, and in particular the nineteenth-century bourgeois house, which the "ruling class" had elevated to the "status of a work of art" ([1932] 2002, p. 164). His tendency to turn away

Figure 3.2. Villa in the Ořechovka garden suburb in Prague. Source: Steven Logan

from individual works of art in favour of the city and its everyday technologies in his writings of the 1920s is mirrored here in his rejection of the single-family house as a work of art by a single creative mind. Although modern architects removed the ornaments, Teige argues, they leave this characteristic of the dwelling unchanged: "a special, isolated object, posing as a work of art" ([1932] 2002, p. 165). It was in this context that he critiqued the star architects who called themselves modern, but who designed villas for the elite. Echoing his critiques of Le Corbusier, in *The Minimum Dwelling*, Teige criticized architects for not living up to the claims of modernism as social cause:

> Instead of holding fast to the principles of economy and functionality and to the promise that one day it will be able to solve the housing problem in the spirit of these principles and on a social scale, architecture has chosen to pander to the rich with a new version of luxury, a luxury of calculated simplicity for their new palaces, posing as modernistic habitable monumental sculptures. ([1932] 2002, p. 182)

Teige criticized Le Corbusier, "who spoke about machines for living and the simplicity of Diogenes's barrel," but who "wastes his time building villas fit for a Midas" ([1932] 2002, p. 182). He called Mies van der Rohe's Tugendhat Villa in Brno one of the seminal works of the International Style and now a very popular museum, the "pinnacle of modernist snobbism" ([1932] 2002, p. 7).

Although Teige associated his critique of the single-family house with the English garden city approach and the ideologies of Ebenezer Howard and Raymond Unwin, he was much closer to Howard in his approach than he would have ever admitted. Indeed, the critiques described above apply less to Howard than they do to Barry Parker and Unwin, the architects of the first garden city, in Letchworth, UK, and the planning principles that developed out of Howard's work after his death. Parker and Unwin draw on traditional architecture – the villages of fourteenth-century England – rather than following Howard's plan for "geometric boulevards and iron-and-steel Crystal Palaces" (Fishman, 1977, p. 69). Teige's claim that "no architectural problem can be separated from its socioeconomic relations" ([1932] 2002, p. 325) mirrors Howard's own vision for the garden city as part of "radical social change" based on cooperative living (Fishman, 1977, p. 62). Although Howard was operating within a capitalist environment, he still sought to wrestle power away from the landlords and towards the community, who would effectively become the holders of land in the garden city, not the state. Occupants would not own their houses, but rather pay rent to a community fund that would go to paying back the original purchase of the garden city land (plus interest), with the rest going towards collective services for the community. This was achieved through Howard's understanding of rent, which he divided into three categories: landlord's rent, sinking fund, and rates, the total of which Howard called "rate-rent."[6] Howard's focus on rent rather than homeownership shows that one of the dominant aspects of the North American suburb – single-family homeownership – was not necessarily the only social-economic model under consideration.

Teige's critique of the garden city approach was made years later by F.J. Osborn in the preface to the 1946 edition of *Garden Cities of To-Morrow*. Osborn argues that the over two million people added to Greater London's population between 1898, when Howard first published his book, and 1945 did not "reduce congestion much, if at all" because the importance of the city centre as a business district only increased and impinged on land formerly used for dwellings ([1946] 1965, p. 14). Robert Fishman notes that Osborn, who himself worked with Howard, gave up on Howard's hope for a "multicentered society," and helped develop "satellite towns," which remained peripheral, physically and symbolically, to the city (1977, p. 84). Howard envisioned both single-family houses and houses with "common gardens" and "co-operative kitchens" ([1946] 1965, p. 54). The very basis of the garden city was the cooperative aspect of the dwellings – namely, the quadrangle, of which three sides would be made up of homes and the fourth side would contain communal facilities: dining, recreation, day care. Cooks would be hired cooperatively by residents. This communal aspect, shared by both Howard and Parker and Unwin, was not realized in practice (Fishman, 1977, pp. 70–1), with Osborn favouring the detached, single-family house as the only environment suitable for a family (Fishman, 1977, p. 84). Howard's garden city was a utopian vision of radical social change, one that called for decentralizing the big cities and creating a network of self-sustaining, cooperatively run garden cities, but they often become less about social change and more about a planning model that offered family homes with a garden away from the industrial zones of the city.[7]

Teige's critique, then, applies less to Howard and more to the way his ideas were taken up, both by his contemporaries and by those who succeeded him in his efforts. Howard's ideas achieved widespread recognition largely after his death, but they persisted as a "planning movement" that would save capitalism, rather than a "social movement" that would help overthrow it (Fishman, 1977, pp. 62, 65). Teige's critique of the single-family house, and particularly modernist architects' penchant for championing social causes while at the same time devoting their artistic and architectural energies to villas, signified a growing theme within socialist urbanism: shifting the aims and goals of modern architecture towards the socialist transformation of the environment as a whole.

The Socialist Green City

Teige's functionalism was matched by calls for a new way of life that he dubbed "poetism," the art of living well. Poetism would break down the barriers between art and everyday life, between art and its singular creator, and between art and technology, and as such was the more poetic companion to Teige's austere functionalism. Poetism was hatched in the cafes of Prague and on night-time walks through the city. Architect Karel Honzík describes the mood on these walks: "We spoke about the future of the world, about the future of creativity. Global revolution was knocking at the door. How would one work, live? How would cities be built? How would one eat? Sit? Travel? Fly?" (1963, p. 68). They would walk "from the moment the street lamps were lit until the moment they were extinguished," traversing the Old Town out to the surrounding districts of Vinohrady and Smíchov (Honzík, 1963, p. 68). Poetism was created in the movement through the city. Poetism is itself defined by and produced in the restless movement of people, ideas, and machines.

It reached its apogee in Teige's call for a "magic-city" (he used the English phrase). In his second poetist manifesto, Teige challenges poetism to organize in the "metropolis of work and production" an "Epicurean garden of poetry, a magnificent and entertaining ... magic-city" ([1928] 2004, p. 89). The "magic-city" would be a city of noise, sounds, colour, and light, a "giant, dynamic symphony" in which movement would be "the fundamental element." It offered a playful vision of Futurism – in fact, in 1921, Futurism's founder, F.T. Marinetti, came to Prague to great fanfare, and Teige frequently wrote about Marinetti and Futurism.

The magic-city is particularly meaningful in Teige's work in this period because through it he connected poetism, socialism, and the city. Teige envisioned the magic-city as a city of encounter, gratification, and a poetry that is "for all the senses" and "never l'art-pour-l'art" ([1930] 2004, p. 236). He also made the connection to Marx's work explicit when he quoted from the *Economic and Philosophical Manuscripts 1844*, in which Marx writes that "the forming of the five senses is a labor of the entire history of the world" (quoted in Teige, [1930] 2004, p. 233).[8] Abolishing private property and the need to own things would mean the "complete liberation of human feelings and qualities" (Marx quoted in Teige, [1930] 2004, p. 234). In this sense, Teige's magic-city can be understood as a form of Marxist humanism and an alternative to the strict functionalism he advocated in his debate with Le Corbusier. Teige's vision was not premised on the ownership of things – cars and houses – but on a mode of aesthetics that called for a deep, sensual engagement with the city and that also rejected the increasing commodification of everyday life. Teige believed that his strict constructivism – the logic of the machine for living – would nourish and feed the development of poetism: together they would create a unique socialist space that Teige described as "green cities" of "deurbanized settlements" in collective dwellings, in a society which has "done away with the institution of the family and freed erotic feelings from their material relations" ([1930] 2004, p. 235).

Whereas the idea of the magic-city is largely found in Teige's writings on poetism in the 1920s, in his introduction to Teige's *Modern Architecture in Czechoslovakia*, Jean-Louis Cohen suggests that Teige's magic-city could easily have been an entry into the 1930 "Green City" competition in Moscow (2000, p. 38). Organized by state labour unions, the competition was for a "socialist garden city" situated outside of Moscow for 100,000 inhabitants (Mumford, 2000, p. 44). It was intended that workers from the city would go there to relax. Le Corbusier based his designs in *The Radiant City* on some of the Green City proposals. CIAM historian Eric Mumford (2000) notes that the competition, which brought designs from many of the Soviet constructivist architects, could be divided into two camps: the disurbanists, who favoured detached dwellings, and the urbanists, who favoured collective dwellings. The distinction is itself muddy, particularly in Teige's writings, as both groups for the most part rejected the existing cities and favoured new settlements. But it would set the tone for debates on socialist urbanism that would continue into the post-war period, specifically the focus on dwelling, mobility, and the rejection of the existing cities.

The disurbanists, of which Mikhail Okhitovich was the leading theoretician, rejected the city and in its place envisioned settlements of mass-produced, single-family homes situated along transportation routes (trains, buses, cars) in "unspoiled natural surroundings" (Kopp, 1970, p. 172; Mumford, 2000, p. 45). Okhitovich's position on the existing cities was clear: "The city must perish … The revolution in transportation and the spread of the automobile will overturn common assumptions about the density and accumulation of buildings and apartments" (quoted in Paperny, 2002, p. 37). Settlements built around the new hydroelectric power stations throughout the USSR would help eliminate the contradiction between urban and rural, city and country (Kopp, 1970, p. 173). The linear cities of the de-urbanists had neither a centre nor a "central business district" (Teige, [1932] 2002, p. 320) and the program in its entirety meant "the end of streets and squares, of the entire classical conception of the urban landscape" (Kopp, 1970, pp. 173–7). The disurbanists rejected the city, which they saw as largely a capitalist beast, and an unruly one at that – in their view, Moscow was to become an "enormous garden of culture and rest" with the clusters of communal dwellings built among the forests of Moscow's periphery (Kopp, 1970, p. 179).

Deurbanism put its hopes in the technological possibilities of mobility, reflecting the "unconditional readiness for mobilization" that Walter Benjamin observed in Moscow ([1927] 1999, p. 29): even the houses should be mobile, prefabricated units, so that if necessary, people could pack up and move them.[9] This was a defiant gesture against the seemingly anachronistic life of the bourgeois who, claimed architect Alexander Pasternak, was "chained to his house" and "no longer an active participation in a fast-moving and changing life" (quoted in Kopp, 1970, p. 177). Teige himself asked: "Why should a dwelling, which is much like a suitcase accompanying our life's journey, be dragged along like a heavy burden?" ([1932] 2002, p. 351).

The basis of the urbanist approach was the communal dwelling, or the *dom-kommuna*, over the single-family house and for which Teige strongly advocated in *The Minimum Dwelling*. The minimum dwelling at its core implied the dissolution of the bourgeois family and its privatized domestic spaces, the liberation of the housewife, and the privileging of communal activities, in particular child-rearing, as the children would live in adjacent "children's homes." Prefiguring the megastructure idea of everything under one roof, the "nerve centre and heart" of the communal house were not the individual dwellings but the collective spaces: auditoriums for lectures, screenings, and theatre, exercise rooms, library, and children's living quarters (Teige, [1932] 2002, p. 385). It would offer everyone a minimal amount of private space but a maximum amount of public space.

These ideas proved influential on Teige as well as the next generation of architects he would influence. In *The Minimum Dwelling* ([1932] 2002), he details the Soviet approach to the communal house, but also focuses on the Czechoslovak architects who were building small apartments and proposing collective dwellings, like Josef Havlíček and Karel Honzík, Jan Gillar, Josef Špalek, J.K. Říha, Ladislav Žák, and the PAS group (made up of the architects Jiří Voženílek, Karel Janů, and Jiří Štursa).[10] Teige was adamant that architects not simply design small-sized apartments, but that they take the everyday life of the proletariat as their guide – following from the Soviet and Czech influences, this meant the elimination of the kitchen, freeing women from the drudgery of housework and cooking, and designing apartments for a single person, rather than small-sized apartments where multiple families were forced to live together ([1932] 2002, p. 389). For Teige, the greatest opportunity to build collective dwellings in Prague came with the VČELA workers' cooperative competiton in 1931 to build minimum dwellings in the Prague district of Vršovice. All of the above-named architects participated and Teige showed and discussed their designs in the concluding pages of *The Minimum Dwelling*. But by 1932, when the book was published, it was already clear, to Teige at least, that the collective apartment designs would not be embraced by the cooperative, and due to a number of other obstacles, the designs were never realized. The idea of dissolving the family unit was also already falling out of favour with the Soviet regime under Josef Stalin (Guzik, 2009, p. 404).

By the mid-1930s Teige became increasingly disillusioned with the Soviet Union following Stalin's embrace of neoclassicism for the Palace of the Soviets and reports from Krejcar, who had been working in Moscow with Moses Ginzburg in 1934–5 (Cohen, 2000, p. 19). In *Vyvoj Sovětské architektury* ([1936] 1969), Teige claimed that the ideals of Soviet architecture no longer emerged out of "the fantastic dreams of the new, free man," but rather out of the ideas of "stuffy Soviet academics and bureaucrats" (p. 77). By the time this book was first published, Teige's key role in the architectural avant-garde was coming to an end (Spechtenhauser & Weiss, 1999, p. 241); however, his ideas lived on among the architects mentioned above, particularly through the PAS group, whose members helped realize Teige's calls for standardization and industrialization of building in the post-1948 period, and the work of architect Ladislav Žák, who continued to advocate for collective dwellings (Švácha, 2000, pp. 95–6).

Teige's final work on architecture and urbanism was the extended introduction he wrote to Žák's 1947 book *Obytná krajina* (*Inhabitable Landscape*). In it, Teige argued that the machine for living, the starting point of architecture and urbanism, had turned into a "dwelling for a machine, a mechanized human" ([1947] 1994, p. 289). The very same technologies Teige once praised had turned dwellings and cities into "the 'world we live in,' but where it is impossible to live" ([1947] 1994, p. 283), a world ravaged by both war and the expansion of the capitalist and socialist economies.

In the same year, Theodore Adorno and Max Horkheimer published the *Dialectic of Enlightenment*. The opening paragraphs of the essay on "The Culture Industry" express a similar disillusion with architecture. They write that the urban planning projects that were to give individuals their own spaces in "hygienic small apartments" have turned into "dismal ... residential blocks" that "subjugate them ... more completely to their adversary, the total power of capital" (Adorno & Horkheimer, [1947] 2002, p. 94).[11]

As attractive as their pessimism might be, Adorno and Horkheimer were in all likelihood unfamiliar with what was happening on the ground in the socialist countries. The early post-war period in Prague did see attempts to put into place ideas around collective dwellings, even if the dissolution of the family was no longer a part of it, freeing women from mundane household tasks and offering a rich, communal environment, evident in the early post-war suburb of Solidarita (explored below). And two collective dwellings were built in the post-war period by architects influenced by Teige: Jiří Voženílek's apartment building in Zlín in 1947 (shown in figure 3.3), and Evžen Linhart and Václav Hlinský's development in Horní Litvínov (construction began in 1946 and did not finish until 1958). Although discussions

Figure 3.3. Collective apartment building in Zlín. Source: Steven Logan

about the minimum dwelling no longer took the socialist-utopian form of the 1920s, the rejection of the ownership of things that Teige had long propagated – using the figure of Diogenes and his barrel as his frequent reference – also survived in the post-war critiques of capitalism, industrialization, and consumerism, along with a revived socialist humanism.[12]

Early Post-war Suburbs: Solidarita and Poruba

Post-war socialist suburbanism, argues Hirt and Kovachev (2015), was dominated by the state, which had a monopoly when it came to constructing new settlements. Their argument focuses on the massive and monotonous developments of the 1960s and '70s, as these places dominated the urban periphery in the post-war period. During the late socialist period, these developments were virtually the only form of building occurring, the legacy of which can be seen in the number of people living in apartment blocks constructed in this time (as of 2009), from Bucharest (82%) to Prague (32%) (Hirt & Kovachev, 2015, p. 193).

Jiří Musil (1985) points to two overriding influences in the building of the Czech sídliště: the architectural avant-garde of the interwar period (see also Zarecor, 2011, pp. 16–21), and the politics of socialist housing, which grew out of the discussions, debates, and architecture of minimum dwellings. Together, these would come to define both the form and content of the sídliště. Musil lists a number of characteristics that define socialist housing politics (1985, pp. 36–9). Harkening back to the socialist minimum dwelling, Musil explains that socialist housing was aimed at eliminating the housing shortage by addressing the needs of the most socially disadvantaged groups and improving the housing conditions overall for the working classes through such measures as decommodifying housing, stabilizing rent so that it is affordable to everyone, and through the industrialization and typification of all the building elements. The approach to socialist housing was not just about providing affordable, mass-produced apartments, it also emphasized a "complex concept of dwelling," which included basic amenities for everyday life, as well as recreational, cultural, and social facilities – in the 1960s this would become an increasingly important aspect in discussions of the architectural living environment (životní prostředí). This was the thrust of architect Karel Honzík's claim in 1953 that "architectural production during Socialism is not constituted from the design of singular buildings, but becomes the production of human environments" (quoted in Miljački, 2017, p. 129). The focus on social and public life emphasizes the "collective form of life" that was a necessary part of socialist ideology.

Post-war socialist suburbanization, particularly in Czechoslovakia, exhibited considerable diversity in the period between 1945 and 1975, even if the trend was towards growing standardization (Musil, 1985, pp. 39–40). Musil suggests three periods in the development of post-war urbanization: the very early post-war period marked by smaller developments that continued the functionalist principles of the inter-war years; a second period marked by the building of socialist new towns in the style of socialist realism; and the final stage, which was characterized by massive developments like South City. The development of the socialist suburb in the 1945–75 period offers a productive comparison to what Canadian suburban historian Richard Harris (2004) describes as Canada's "creeping conformity." Although Harris's history stretches back to 1900 and includes working-class and self-built suburbs, this creeping conformity refers to the post-war suburbs built by large land developers and made affordable through debt. By 1960, writes Harris, "they were more uniform and ubiquitous than ever and were leading symbols of a new consumer lifestyle" (2004, p. 164). The post-war sídliště, without private developers or debt financing, exhibited its own uniformity and ubiquity and, especially in the 1960s, struggled with accommodating the growing increase in free time.

This, however, was not the case with two early post-war developments that Musil mentions – Solidarita in Prague and Poruba in Ostrava, both of which belong to the early phases in sídliště building. Situated on the periphery of the city (at the time of their contruction), these developments are part of the modernist and socialist urbanisms and make interesting use of the relationship between the street and the dwelling.[13] They do not simply defy the stereotype of the mass-produced suburbs being constructed at the same time in North America, they also bring an added depth to the idea of suburban superblocks. Like their counterparts in the West, they are notable for the particular attention paid to the pedestrian landscape and common garden spaces. Although Poruba is not in the Prague region, it is one of the best examples of a socialist suburb of the socialist realist period.

Solidarita (1947–9) is made up primarily of small, single-family row houses, although the 1,200-unit complex also includes apartments (see figure 3.4). Musil notes that the early suburbs like Solidarita "represented a continuation of the inter-war functionalist concepts" (1985, p. 39). This was particularly evident in Solidarita's attempt to institute some of the cooperative and communal principles that were discussed in the 1920s without wholly embracing Teige's approach, and in particular the dissolution of the family, which in the post-war socialist period was seen as an extreme form of collectivization (Guzik, 2009, p. 407). Solidarita combines some aspects of collective living with the garden city principles of the 1920s, and as such, signals a tension in both the interwar years and the post-war years between the single-family house and the collective dwelling as the ideal of socialist living.[14]

Solidarita is also important because it marks the beginning of a long-standing influence of Scandinavian architecture and urbanism on post-war developments, from collective dwellings in the 1940s to the new towns of the 1950s and '60s. The housing cooperatives in Scandinavia, writes architectural historian Hubert Guzik, offered a "successful combination of individual and collective elements in housing construction and … a model for overcoming social prejudice against collective accommodation" (2009, p. 405). Guzik (2013, p. 43) also notes that during the time of the occupation during the Second World War, architects and urbanists became familiar with Scandinavian architecture through international magazines and journals available at the time; although this was not the first time that Scandinavian architecture was featured in Czech-language publications. Europe's first apartment building containing a central kitchen was built in Copenhagen in 1903 and subsequently featured in a 1908 edition of Brno's *Lidové noviny* newspaper

Figure 3.4. Aerial view of Solidarita. Source: *1000 let stavby Prahy* (Borovička & Hrůza, 1983)

(Guzik, 2012, p. 47). Complicating the line of influences, Otto Fick, the architect, cited American apartment cooperatives as one of his influences. In 1946, the architects of the two collective apartments mentioned above made two trips to Sweden to look at the country's collective apartments (Guzik, 2012, p. 47). The architects of Solidarita specifically drew on a Scandinavian cooperative housing project in Praesthaven, Denmark (Zarecor, 2011, p. 268).

Solidarita, in typical modernist fashion, alters the relationship between the street and the dwelling. As is common in early post-war socialist housing, the rows of houses run perpendicular to the main road, but there is no vehicle access to the front of the houses, which are accessed by pedestrian walkways running between the front- and back-yards of the houses (shown in figure 3.5). Their backs face the common gardens (although in some cases, fences have been erected between plots). Here the superblock comprises seven rows of housing, with only pedestrian streets running between the rows. The result is not a superblock as it is usually thought of, but a network of sidewalks in a lush landscape amidst very modest-sized row housing.

Poruba, begun in 1951 and planned initially for 150,000 residents, has both a monumental scale, displayed in its wide avenues, and a human scale, which we see in the intimate spaces of the kvartaly. The buildings lining the main thoroughfare, Lenin Street, had shops and

Figure 3.5. Pedestrian pathway leading to the front doors of row housing in Solidarita. Source: Steven Logan

restaurants on the ground level and apartments above, and today a bike path that runs down the centre of the boulevard is surrounded by rows of trees on either side (see figure 3.6). Poruba's apartments were arranged in ensembles around courtyards, accessed via gateways through which pedestrians and cars would enter from the street (see figure 3.7). Zarecor notes that it was influenced by the superblocks of the Vienna apartments of the 1920s. In contrast to the tower in the park ideal, the superblocks or building ensembles were entered through elaborately decorated openings that connected the street to semi-private courtyards. These blocks often were as large or larger than a city block (Zarecor, 2011, p. 173). Historian Catherine Cooke shows how the ensemble had a meaning beyond its strict architectural and urbanistic one: offering a series of "ever-larger ensembles" that began with the "harmoniously composed building ... and extended out into the city and conceptually to the whole socialist world" (1997, p. 149). The gardens are now beautiful green spaces where people can gather; this is nothing like the stereotypical depiction of the so-called tower in the

Figure 3.6. Hlavní třida boulevard in Poruba. Source: Steven Logan

Figure 3.7. Gateway leading into one of Poruba's superblocks. Source: Steven Logan

park – a tower surrounded by what is often depicted as a bare, unwelcoming landscape. In many ways, like Solidarita, they offer a parallel to Stein's superblocks in Sunnyside Gardens and Radburn without, however, the demands of the automobile, which was not a deciding factor in the building of these early post-war suburbs. Some of the building entrances were within the courtyard itself, rather than the street side, similar to the apartment complex in Sunnyside Gardens.

Where the facade might have disappeared under modernism, it makes its grand reappearance in Poruba, part of the delineation of private and public space, of street space and dwelling space, but also part of socialist realism's *Gesamtkunstwerk*, or total work of art (Cooke, 1997, p. 148). Poruba's facades included elaborate *sgraffito*, pictures on the walls of buildings depicting peasant and factory worker scenes. As Cooke writes, "A confident, radiant society reflected its certainties through *obrazy* [images; pictures] of clear form and clearly understood meaning" (1997, p. 143). The defining monumental element of Poruba was the *oblouk* (archway), which functioned as the gate to Poruba, the most monumental of the street-spanning buildings, with the tram running underneath the apartments (see figure 3.8).

Poruba was a distinct suburb of Ostrava situated far from the mining and industrial sites. The minister of heavy industry, Gustav Kliment described it as "pretty land, in the picturesque countryside, surrounded by forests," which allowed miners "who spend all day digging out coal without a ray of sunshine, to at least have enough sunlight in the hours that they have for rest" (quoted in Zarecor, 2011, p. 176). Poruba as a work of socialist realism draws on the pre-capitalist traditions of spaciousness, greenery, and low density, which were the hallmark of the Soviet proletarian city going back to the 1935 Moscow Plan (Cooke, 1997, p. 155). The courtyards interweave green spaces and dwelling spaces and the streetscape has not been sacrificed, even though there are clear attempts to separate cars and pedestrians, both on the main thoroughfares and within the courtyards.

Solidarita and Poruba offer two distinct alternatives to the neighbourhood unit. Poruba is revered for its green spaces, and Solidarita for its modest neighbourhood unit, one that offers small, single-family dwellings and apartments, developed around collective facilities and built at the scale of the pedestrian, not the car. The neighbourhood unit of Solidarita and the kvartaly in Poruba, offer a good comparison to the early post-war attempts to implement the neighbourhood unit in Toronto, such as Don Mills (discussed in chapter 5), where the emphasis was on single-family homes and the road network in which the four neighbourhoods were situated.

Figure 3.8. The oblouk in Poruba. Source: Steven Logan

The Socialist Suburb after Khrushchev

In 1956, two years after his speech criticizing socialist realism and encouraging architects to learn from the West, Khrushchev claimed that the Soviet Union would not follow the lead of the United States in increasing private automobile production and promoting individual consumption. Khrushchev called the standard (and sacred) practice of private cars and drivers for Communist state officials "wasteful" (Siegelbaum, 2008, p. 224); instead, he wanted better public transportation and networks of affordable taxi fleets (Gronow & Zhuravlev, 2010, p. 134). Here is how he described such a system to a Paris audience in 1960:

> Under such a system of car use we will obviously need 10–15 times less cars in comparison with the case where the ambition would be to provide everybody with a car. We will meet people's needs for [a] car in a more rational manner. After all, people are not tramps. People work, and while they work, the car is parked. While it is parked, it does not do any good, although its age increases just as much. We would regard it, therefore, as quite irrational to allow a wasteful increase in the number of cars. For us, the capitalist, private ownership-based use of the car is a path to be

avoided. We will provide [for] our population in a socialist manner. (quoted in Péteri, 2009, p. 8)

Khrushchev's speeches set the stage for large-scale socialist suburbanization beginning in the late 1950s, which was based on the standardization and typification of dwellings – of which the prefabricated concrete panel apartment building was the quintessential product – but also on the creation of a distinct socialist living environment that would privilege the pedestrian, public transportation, and modest apartment living, over the cars and sprawling suburban houses of the West. In particular, his 1954 speech criticizing socialist realism and extolling the virtues of mass production signalled in part a return to the ideas of the interwar avant-garde on the machine for living, architecture-as-instrument, and the industrialization of building. The speech is key to explaining the dominance of prefabricated building technologies in the massive housing developments of the 1960s and '70s, which form the core of the second act of Hirt and Kovachev's (2015) story on Eastern European suburbanization.

The post-war story, as the above examples already show, is more complicated. At their annual meeting in 1956, the Union of Czechoslovak Architects (Svaz architektů ČSR) rejected the "formalist excesses of historicism" that was associated with socialist realism, but at the same time cautioned against "a mechanical embracing of functionalism," prefiguring the later reworking of Athens Charter urbanism called for by architects and planners in the 1960s (quoted in Musil, 1985, p. 40).

Khrushchev's rejection of socialist realism actually became a catalyst for the increased autonomy of architects and urbanists from state dictates. According to Cathleen Guistino (2012), historian of Central and Eastern Europe, the ensuing decades are less indicative of an easy story of totalitarian states dictating to architects what to build than of "the deficiencies of the old totalitarian paradigm resting on suppositions about monolithic, omnipotent states in which populations had absolutely no autonomy" (pp. 189–90). Ana Miljački (2017) similarly argues that although architects needed to fulfill the state building quotas, especially in the 1960s, they "received only basic ideological guidance from party officials" (p. 183). They increasingly turned to their colleagues in the West because they shared many of the same post-war challenges and shared their misgivings with the functionalist urbanism of the interwar period as a solution. Architects on both sides of the divide were addressing increased leisure time.

The wider atmosphere of international cooperation was marked by the signing of the 1958 US-Soviet Cultural Agreement and Expo 58 in Brussels, the first world exposition of the Cold War. The very successful Czech pavilion set in motion a "Brussels style" that would go on to influence architecture and design in Czechoslovakia throughout the 1960s, while offering a new image of socialism. The successful pavilion draws out a number of elements that Miljački (2017) associates with post-war Czechoslovak architecture: teamwork, synthesis, and standardization of building as the defining element of the state economy. Teamwork spoke to the architectural profession and the organization of architects into collectives, while synthesis referred to the ways in which teams of experts worked together. As Miljački herself shows, these terms had their origins in the interwar avant-garde, and so illustrate continuity between the time periods. In *Vision in Motion* ([1947] 1965), László Moholy-Nagy critiques the specialists of industrial society who work at their individual tasks while "missing both human and social direction" (p. 16). In order to undertake such analyses, Moholy-Nagy calls for an "international cultural working assembly" of "scientists, sociologists, artists, writers, musicians, technicians and craftsmen" devoted to problems as diverse as urban planning, nutrition, production and dwelling, media, folklore, crime, etc. In the final sentence of *Vision in Motion*, Moholy-Nagy writes that this assembly could "translate Utopia into action" ([1947] 1965, p. 361).

Teamwork, synthesis, and standardization were central to the discussions of the socialist living environment. Miljački writes that the 1964 meeting of the Union of Czechoslovak Architects, which took place the same year in which the plans for Prague's three biggest sídliště were unveiled, including South City, reaffirmed the importance of a "synthesis of the work of architects and artists within an urban dimension" (2017, p. 195). These discussions began with Expo 58 and reached their high point with the 1967 UIA meeting in Prague.

Czech architecture and urban theorists associated with the *Panelací* project have since revised Musil's (1985) three stages of the sídliště to account for the diversity in building in the wake of Expo 58.[15] Whereas Musil, like other theorists, put all socialist suburban building after Khrushchev's speech into one group, these theorists delineated four periods from around the time of Khrushchev's speech until the end of Communism in 1989. The years between Expo 58 and Expo 67 are seen respectively as the "pioneering phase," so named for its experimentation with new building methods and typologies, and the "beautiful" or "humanistic" phase, marked by the attempted humanization of the sídliště: a greater attention to the *quality* of the living environment, and not just the *quantity* of apartments. Prague's experimental inner-city

Figure 3.9. Aerial view of Invalidovna, with district centre and hotel apartment towards the back of the photo. Source: *Výstavba hlavního města Prahy, 1945–1975* (courtesy of Miroslava Fišarová)

Sídliště Invalidovna (shown in figure 3.9) is a key example of the early pioneering phase, while two notable examples of the later beautiful phase are Brno-Lesná in the city of Brno (figure 3.10) and Ďáblice (figure 3.11) in Prague.

In 1967, Viktor Rudiš the main architect for the Brno-Lesná sídliště travelled to Paris to work in the studio of Team 10 member Candilis-Josic-Woods (Hanáčková, 2014, p. 83). Describing a visit to Otaniemi, Finland in 1968, the Czech sociologist Bohuslav Blažek later remarked on how "people could leave their houses on skis and enter into a forest, as the buildings gave way to fully-grown trees" (Blažek, 1998, p. 41). Tapiola near Otaniemi was also the model for Brno-Lesná, which Rudiš drew on, especially the idea of dwellings situated in a forest-like setting. Here is how Rudiš described his visit to Tapiola: "We finally saw

Figure 3.10. Sídliště Lesná in Brno. Source: Martin Horáček

with our own eyes dwellings in greenery, in beautiful natural surroundings, forests, water parks, rocks and, between it all, modern buildings" (Chatrný, n.d.). In her history of new towns, Rosemary Wakeman places Tapiola, along with Vällingby, at the pinnacle of Scandinavian social democracy. And as Blažek noted, the planners explicitly evoked a life lived in nature, even though Tapiola "pioneered a modern lifestyle devoted to recreation and consumerism" (Wakeman, 2016, p. 99), and thus at first glance would seem at odds with socialist principles. And yet this tension was central to the socialism of the 1960s and onwards, which sought to reconcile mobility, new technologies, and the increasing importance of leisure with the tenets of socialism. In the socialist suburbs, the production of nature, leisure, and recreation were paramount, but without the consumerism. Rudiš would later be critical of his work in Brno-Lesná for its lack of urbanity, and in some ways the plans for South City sought to combine Brno-Lesná's emphasis on living in nature with the urbanism of a city centre.

The inner-city development of Invalidovna was explicitly billed as an experiment in sídliště building. It meant to open up the possibilities for new types of construction and to develop a new kind of neighbourhood

Figure 3.11. The district centre of the Ďáblice sídliště. Source: *Výstavba Sídlišť* (courtesy of Miroslava Fišarová)

unit and new forms of dwelling – in Invalidovna's case, this included a "hotel apartment" that would cater to people living alone, or with a small family (the apartments were furnished with a little kitchen nook; see Guzik, 2014, p. 64). Invalidovna also experimented with creating new kinds of collective spaces. Instead of having all the shops in one building or in the ground floors of the apartments, Invalidovna featured an outdoor "district shopping centre," easily accessible from the road or from the apartments on foot (unlike Perry's neighbourhood unit, where the shops were situated at the intersection of two busy roads). Invalidovna's centre (pictured in figure 3.12) was scaled to the pedestrian, and featured a range of amenities (the first plans were designed by South City's chief architect, Jiří Lasovský, but the project was, in the end, given to a different architect, Milan Rejchl). In many ways, Invalidovna set the tone for the more experimental projects later in the decade, like Etarea and South City.

Figure 3.12. Invalidovna district centre. Source: *Zdeněk Vořenílek* (courtesy of Ladislav Zikmund-Lender)

Figure 3.13. Crane urbanism in post-war Prague. Source: *Zdeněk Vořenílek*

Crane Urbanism

The lasting and dominant image of the post-war socialist suburb is the row upon row of prefabricated apartment blocks in a landscape wiped clean by their construction. Like Invalidovna, developments were intended to be self-sufficient, with not only the most necessary social services – like hospitals, clinics, and schools – but also shops, cafes, cinemas, sport facilities, etc. However, Sonia Hirt (2012) notes that because of the lack of resources, most of these latter services were not built, leading Sofia's resident to call these large dwelling complexes "un-complex complexes" (p. 87) – that is, although like their socialist-realist predecessors they were to be harmoniously composed ensembles, they ended up as a collection of individual residential buildings within an unfinished, and unforgiving, landscape. Hirt notes that the socialist cities often suffer from a lack of small retail and "in-between" public spaces like cafes, small shops, or yards (2012, p. 40), and not just overt, grand public spaces like "friendship parks," large cultural centres, etc.

Socialist cities were unique not only because they were planned but also because their ideologues claimed that those plans always translated into reality (Zarecor, 2011, p. 152). This may have been the case with Poruba, but the later developments differed in many ways, and it is these later developments, simply because of their sheer size and the number of people who lived in them (and still do), that represent the lasting legacy of socialist urbanism. In order to take advantage of economies of scale, developments had to be larger. The construction companies and state investors wanted projects of 5,000 to 15,000 apartments and not "the smaller housing projects of low-rise, detached buildings in order reap (supposed) economies of scale" (Szelényi, 1996, p. 305). Jiří Hrůza describes the situation in Prague in the 1960s, where even though planners insisted that maintenance and reconstruction of the existing building stock was necessary, "we had to fulfill the demand: find land for around 11,000 apartments annually" (2006, p. 38). The result was dilapidated housing in the older, more central neighbourhoods (Szelényi, 1996, p. 305).

The whole building apparatus was geared to such development: the prefabricated panels were constructed off-site in "house factories" (Szelényi, 1996, p. 305) and transported at great expense, both financial and infrastructural, due to the wear and tear done to the unprepared road system. Once at the construction site, workers had to assemble the panels into apartments with the help of cranes, which themselves required vast swaths of open space to operate. All construction was geared towards enabling this technology, which, because it was expensive, lead to the shortcuts on "unnecessary" construction of cultural facilities, shops, etc.

The privileged actor in the industrial construction of these housing developments was neither the worker, reduced to stacking boxes one on top of the other, nor the architect, sidelined by the construction: it was, rather, the crane, a feature across all socialist suburbs. Czech architects even came up with a moniker for this kind of building: crane urbanism. Hrůza describes crane urbanism as follows: "For them [the state building organization] it was easiest if everything was made with one kind of panel, brought to the site and assembled" (2006, p. 38). Moreover, the crane could not function without the rail tracks upon which it moved. Hrůza adds: "it was neither cheap nor easy to build the track for the crane, so once it was there, they wanted to assemble around it as many apartments as possible" (2006, p. 38). And the longer the building the better – the technology, ironically, enabled the realization of the avant-garde's dream of "kilometre-long rows of housing" (Teige, [1932] 2002, p. 320). Crane urbanism "seriously undermined the ability of architects and planners to actually *design* places" (Hirt, 2012, p. 39), partially because the expense of building this way necessitated that the projects economize "at any costs" and the victims were usually the urban plans' architectural details (Maier et al., 1998, p. 55). Not only did the placing of the crane tracks help determine the arrangement of the buildings – much to the horror of the architects – but the only place where there was enough room for the tracks, and where the economies of scale of crane urbanism and the state building quotas could be met, was on the city's periphery. Crane urbanism, then, might be better thought of as crane *suburbanism.*

In an edited collection on the culture of dwelling in the 1970s, one contributor notes: "In the end it was rather the tracks of the crane, transporting the prefabricated panels around the construction site, that determined the architecture and arrangement of 'mass housing'" (Koukalová, 2007, p. 75). In neglecting the creative impulses of the architects and the planners, crane urbanism is the ultimate expression of the anonymous mass production that characterizes twentieth-century industrialization, aptly named by Giedion *Mechanization Takes Command* (1948). Hrůza recalls that it was not necessarily political pressure that forced architects and urbanists to build solely with the prefabricated panels: "in reality, the worst lobby was the state building organization – our biggest partner and the biggest antagonist" (2006, p. 38). The "housing construction 'machine'" had become a "sacred cow of social policy" in Czechoslovakia (Maier et al., 1998, p. 57).

Most scholars and architects attribute the socialist suburb's unfinished character to the state's maniacal focus on building apartments as efficiently and quickly as possible. The birth of the post-war socialist city, particularly the third stage of the post-war suburb, takes place in a space already coded by technologies of industrialization and state power, and it expresses a tension between the city to be born and the technologies that would induce labour, to bring about the birth as quickly as possible. Crane urbanism is a major aspect of what Katherine Lebow, in her work on new towns in Poland, calls "state socialism's promised shortcut to modernity" (2013, p. 4).

In the confluence of industrialization and urbanization, crane urbanism and the Athens Charter, the socialist city becomes "a product *strictu sensu*: it is reproducible and it is the result of repetitive actions" (Lefebvre, [1974] 1991, p. 75). For Lefebvre, a product in its strict economic sense is something produced on an assembly line, the result of a rational production process that has as its goal the mass production of any number of identical objects formed out of the "repetitive acts and gestures" of labour and machines (Lefebvre, [1974] 1991, p. 70). The product is not a totality, but a result of conceiving space as both homogeneous and fragmented.

Although crane urbanism has become synonymous with state control over house building in the Soviet-influenced regions, it was not a uniquely socialist phenomenon, and the existence of peripheral urban developments lacking in basic infrastructure and social necessities was also a problem for the cities of the West, and it continues to be a problem on the global suburban landscape. On the ecological implications of crane urbanism, even the much-lauded post-war urbanism of a city like Stockholm suffered under the weight of the crane. Here is how urban historian Stephen V. Ward describes the practice:

> By 1970 construction had been more completely industrialised, with extensive use of low- and high-rise apartment blocks using prefabrication and many residential layouts conditioned by the operational requirements of the construction crane. The way the built environment was placed in earlier schemes within the undulating, rocky topography of Stockholm and its surroundings gave way to sites levelled before construction so that tracks could be laid for the cranes. (2016, p. 321)

The sheer dominance of this type of construction in the 1970s and '80s assured that the unfinished character of the post-war socialist suburb was far more dominant than in the capitalist case. In comparing socialist and capitalist modernist urbanism, Zarecor points out that although

Le Corbusier and Robert Moses have achieved near legendary status in the modernist transformation of the urban environment, neither of them "achieved the scale of urban transformation that one saw in state socialist countries, neither in their own time, nor in the fulfillment of long-term plans that projected their ideas into the future" (2018, p. 5).

Yet in spite of the ubiquity and dominance of the mass-produced sídliště, the tension between industrialization and the living environment did not simply end with the triumph of crane urbanism. The question of the living environment continued to be debated among socialist architects in the pages of architectural publications and the developments in the late 1970s and '80s, which the Panelací theorists refer to as the "technocratic phase" and the "late beautiful and post-modern phase," the latter of which attempted to account for the deficiencies of crane urbanism. Krivý (2017) writes that the sídliště built in the late 1970s and '80s not only drew on postmodern currents, but also revived the socialist realism of the 1950s and in particular the human scale of the dwellings, the squares, and the courtyards (p. 312). Together, they promised "freedom from" the strictures of crane urbanism, while also reaffirming the "universal right to housing" that the socialist city represented, coupled with "the right to a high-quality living environment" (Krivý, 2017, p. 95). The "late beautiful" phase meant the revival of the pedestrian street and a renewed emphasis on the urban *parter* (Krivý, 2016, p. 92). "Parter" is a difficult word to translate, as it can refer to the space between buildings, ground-level storefronts, street furniture, water fountains – that is, the life *between* buildings, rather than the life *in* buildings. In an article specifically addressing the parter, Jiří Lasovský (1981) wrote that people usually pass the way from the bus stop to their apartment as if these are simply "transit corridors." The walkways in the sídliště needed something to attract people's attention, something to draw them out of their apartments and to enliven the space between buildings. Architects believed the parter were most neglected in the "technocratic phase" of sídliště construction. In his article, Lasovský connected the absence of the parter in the new sídliště directly to the crane: the sídliště was great for the tracks of the crane, but not for the "life of the people who would live in this environment for a generation to come" (1981, p. 248). Lasovský's writings on the parter call to mind the way Jane Jacobs praised the street-level elements in Gruen's plans for Fort Worth. The sídliště most evocative of this period is Southwest City, which was part of the 1964 Prague urban plan, along with North City and South City, the architecture of which, Krivý (2016) writes, did not reject industrialization, but rather sought to reform it (p. 96). A further example is South City II-West, discussed at the end of the next chapter.

Conclusion

The roots of the socialist suburb in Czechoslovakia go back to Teige's minimum dwelling and coalesce in the post-war period around the entire environment for living and crane urbanism. A focus solely on crane urbanism plays into the standard narrative of the socialist city, a product of the monopolistic state apparatus imposing its will on architects, planners, and inhabitants. However, a closer investigation reveals a more complex socialist urbanism, one that sought to reform the suburb at the same time that it was subject to the dictates of the state. The South City development is remarkable in this regard. It crosses over into the territory of many of the phases described here, at once pioneering, experimental, and brutally technocratic, in theory and practice, having begun with plans for an experimental suburb, called Etarea, featured at the Czechoslovak pavilion at Expo 67 in Montreal.

4 South City as a Work of Art in the Age of Mass-Produced Dwellings

In the "Conflicts" section of the popular Czechoslovak pavilion at Expo 67 sat a large relief model of Etarea, an experimental suburb for 130,000 people to be situated in the verdant landscape 10 kilometres south of Prague. It was designed by the Pražský projektový ústav (Prague Design Institute, or PPÚ), which also developed Ďáblice and Invalidovna; inspired by its experimental approach, Etarea would try to reproduce Invalidovna on a large scale and on the periphery of the city. According to planners, Etarea was to alleviate the "anxieties of the modern age" and provide an alternative to the disconcerting "anonymous, amorphous metropolis" of the capitalist city (Čelechovský et al., 1967, p. 401).

Expo 67, along with Etarea, was an homage to modern technology and to socialist ingenuity when it came to mitigating the effects of industrial society. Etarea was not just a typical modern city – it was an attempt to form a specifically socialist living environment that focused on cultured leisure and the socialist appropriation of technology. The technological development and general economic growth of the 1960s had made possible the shortening of the work week and the raising of the standard of living. The development of technology in particular brought the possibility of efficient transportation and new building technologies (Hrůza, 1962, p. 11).

The basic philosophy of Etarea, a variation on the communist idea of collapsing the differences between city and countryside, was to connect recreation and housing. Recreational spaces – forests, an artificial lake, and a beach – would be close to dwellings so people would not need to flee the city on weekends for their *chaty* (weekend homes). It was in part a response to the weekly exodus by car from what for many was the alienating and uninviting environment of mass-produced apartments in the sídliště. Ladislav Žák, the author of *Obytná krajina* (*The Inhabitable*

Figure 4.1. The sídliště as a work of art in the age of mass reproduction.
Source: Steven Logan

Landscape, 1947), warned of "recreation transportation" by automobile and the "growing flood of cars and motorcycles dangerously overtaking city streets, highways, and country roads" (1967, p. 4). Although Etarea's designers claimed that "the majority of people's interests could be satisfied right in the city," they still counted on one car for every three inhabitants – that is, one car per family (Čelechovský et al., 1967, p. 400).

Like Invalidovna, Etarea would offer a range of dwellings: high-rise buildings, single-family houses, as well as apartments for individuals (*hotelový dům*). The layout for these small apartments only included a small corner kitchen space with ground-floor collective facilities. Every apartment would have a terrace or a balcony, while ground-floor apartment dwellers would have access to small family gardens. It would expand, though on a much larger scale, Moshe Safdie's "for every house a garden" principle, used in his Habitat 67 prefabricated apartment building built for that year's Expo. The ground floors of the buildings around the centre would also contain shops, cafes, restaurants, and cinemas. It would combine the best of both city and suburb.

Figure 4.2. The "streetscape" in the planned city of Etarea. Source: *Architektura ČSSR, 26*(7) (1967)

Etarea's main form of transportation would be pedestrian, although the development would be well connected to Prague and the rest of the country via its Alweg monorail, highways, and an airport. According to its designers, Etarea would allow for the rapid mobility of vehicles, information, and goods while creating a relaxing, conflict-free suburban environment through the separation of cars and pedestrians. The city would use an elaborate pneumatic tube system with central distribution points to deliver food – ordered via computer – directly to households. The monorail and the main highways would be sunken in transportation corridors.

If residents did not want to push-button shop, they could walk to any one of the 13 local centres, which would be in the form of elaborate, above-ground, pedestrian-only environments, referred to as "habitable street spaces" (shown in figure 4.2). These street spaces would be complemented by an autonomous network of pedestrian pathways amidst green spaces reaching out to the residential areas; on the periphery, the pedestrian pathways would merge with a set of trails and promenades through the forest parks and open spaces. The separated pedestrian spaces were part of this modern utopia of free and easy movement that included a monorail and planned growth of one car for every three residents. Cars would be accommodated in underground "carparks," which would offer direct connections to the largest buildings.

In theory, Etarea addressed many of the post-war issues faced by architects, socialist and otherwise, and it did so in the spirit of teamwork and synthesis that Miljački argues were the key elements of post-war Czech architecture. The team for Etarea featured a range of experts, from Jiří Musil to individuals working on cybernetics and systems engineering, criminologists, medical doctors (including one doctor working on the "psychology of the city"), and geologists, among others.

There was a cover article on Etarea in *Večerní Praha* with the headline "Happy City." Like its Green City counterparts in Moscow, however, Etarea remained a happy city on paper only; it was never realized. But it did become the model for South City, called alternatively the "Sequel to Etarea" (1970) and "Etarea in praxis" by Prague archivist Kateřina Jíšová (2005).

South City was to be the largest housing development in Czechoslovakia: approximately 20,000 dwellings for 80,000 people on 1,200 hectares. The PPÚ was charged with overseeing the design of 25 of the 67 new housing developments in Prague between 1957 and 1985 (see figure 4.3). Of all the housing developments the PPÚ was working on at the time, South City was by far the largest, but also one of the least dense at 66 persons per hectare; for comparison, the PPÚ's most dense sídliště was Invalidovna at 343 people per hectare (PPÚ, 1971, n.p.). The 1964 urban plan included plans for Prague's three biggest developments: Severní město (North City), Jihozápadní město (Southwest City), and South City. These developments signalled a "necessary transition" from the smaller and more isolated sídliště to much larger settlements – that is, "cities" (Borovička & Hrůza, 1983, *p.* 89). In the Czechoslovak state's fifth (1971–5) and sixth (1976–80) five-year plans, the goal was to build over 100,000 apartments (Říha, 2007, p. 21), while in Prague, according to a 1968 interview with Voženílek in *Večerní Praha*, the state planned to build 12,000 apartments yearly from 1971 to 1975. South City, with its 20,000 apartments, would make a significant contribution.

Modelled on Etarea, South City is important as both a modern and a socialist city: its architects responded to the criticisms of the Athens Charter from individuals and groups within CIAM and envisioned a city that would embody the socialist reforms of the 1960s. Lasovský's plans for South City's centre would be the core of the sídliště. It was also a specific response to the problems of the existing socialist suburbs, which Jiří Hrůza called "the crisis of the sídliště."

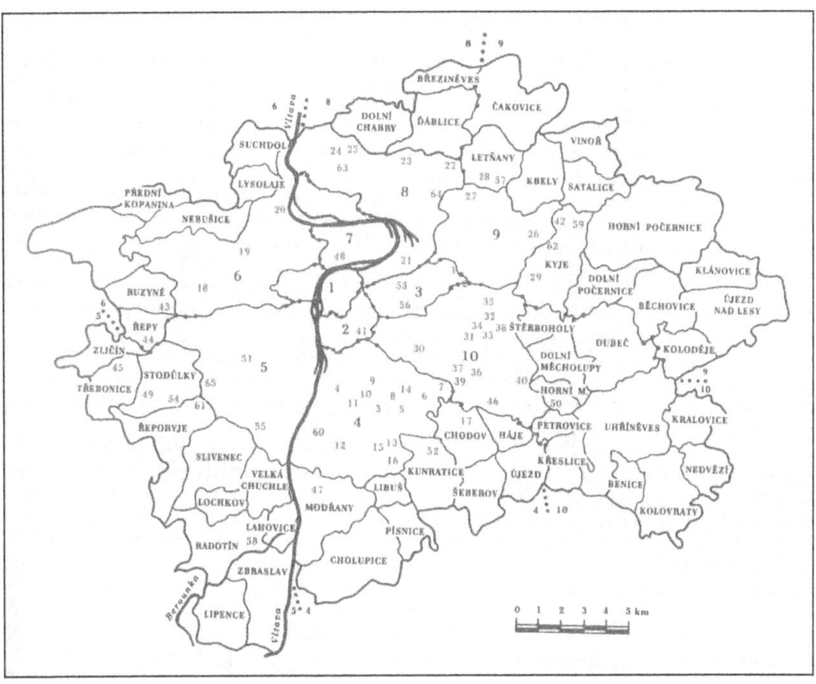

Figure 4.3. Map of greater Prague showing the sídliště built between 1956 and 1986. South City is No. 17 in the southern neighbourhood of Chodov (the larger numbers indicate the city districts). Source: *Výstavba Sídlišť* (courtesy of Miroslava Fišarová)

The Crisis of the Sídliště as a Crisis of Socialism

The crisis, as Hrůza understood it, centred on the living environment of the sídliště, not necessarily the individual apartments. The existing sídliště did not live up to the ideals of the socialist city, certainly as Musil described them in his book. To call the sídliště a suburb is very different from calling Levittown or Willowdale a suburb. The term sídliště has multiple connotations, one of which is something less than urban – namely, suburban, a qualitative definition that differs from the sídliště as suburban simply because it is located on the periphery. Sídliště is a notoriously difficult word to translate. In the Czech online dictionary Slovnik.cz, seventeen different English translations are offered. These include such terms as "housing estate," "housing development," "neighbourhood unit," "commuterland," "commuterdom," and, very simply, "blocks." Although the usual translation is "housing estate," the term does not always resonate with a North American audience, and as such does not capture both its positive and negative connotations, similar to such heavily loaded terms like "suburbia" or the French *grandes ensembles* (the sídliště are also referred to as "large dwelling complexes"). Sídliště is actually an archaeological term that predates its usage in the post-war context; it can refer to any permanent settlement for a group of people and more generally a settlement (*osídlení*) of any kind (Zadražilová, 2007, p. 41). Architect and urban planner Karel Maier calls the sídliště a "mass housing complex," which does not refer only to its post-war incarnation: since the Industrial Revolution, the sídliště have provided a way to house a lot of workers in one place (1998, p. 2). However, the term as it is used today has invariably come to refer to the housing developments comprised almost exclusively of the prefabricated concrete apartment blocks, referred to in Czech as *panelové domy*, or colloquially as *paneláky*.

In his study *Lidé a sídliště* (*People and the sídliště*), Musil (1985) writes that the sídliště are "one of the most distinct markers of socialist-city building and socialist architecture" (p. 13). In his study, Musil compares life in the sídliště with life in the older, inner-city neighbourhoods of a number of different cities in Czechoslovakia. He writes that many of the people who moved to the sídliště did not come from the city but rather from surrounding villages and towns to work in Prague. The majority of the people interviewed were unable to categorize the sídliště in terms of traditional concepts like city, suburb, or small town. The sídliště, he concludes, "is a phenomenon that cannot be classified with the help of old terms" (Musil, 1985, p. 319).

The word is very much similar to the German Siedlung, and takes us back to the 1920s and the "exurban *Siedlung*" at the core of Ernst May's building program in 1920s Frankfurt. Architectural historian Eve Blau's description of May's program resonates with the possibilities that architects and planners held for the sídliště: it "synthesized the *Trabantenprinzip* (principle of the satellite town) with rationalization of production to create exurban *Siedlung* that were 'built utopias,' complete in themselves" (Blau, 1999, p. 170). In avant-garde terms, the Siedlung also became "one of the key sites of typological and technological innovation in architecture in the 1920s" (Blau, 1999, 1999, p. 159). In Vienna, the Siedlung took on a different meaning altogether, as it referred to the self-help cooperative housing where people grew their own food, and which was a product of "radical politics and anarchy" rather than state ideologies of deurbanization and decentralization. The form of the Siedlung was at the centre of debates in both Vienna and cities in

Figure 4.4. South City, 1982. Source: Jaromír Čejka

Germany in the interwar period, and like in the Czech, the German term "Siedlung" generally implied building on the outskirts of the city.

Although the sídliště came under increasing scrutiny in the post-socialist period – in particular from Václav Havel – the critiques had already begun not long after the state began constructing them.

The sídliště were "grey, monotonous, dull, [and] lifeless" (Gottlieb & Todlová, 1969, p. 211), "parasitic cities" (Nový, 1971, n.p.), and "dormitory suburbs" lacking in public, social space (Hrůza, 1967a, p. 1). Such negativity does not simply stem from the ubiquitous nature of the apartment blocks. "By the mid-1970s," notes Krivý, "a widespread dissatisfaction with the ostensible monotony and homogeneity of the sídliště was firmly in place" (2017, p. 305). Dissident and writer Egon Bondy captures something of the spirit of this critique of the sídliště of the post-war period – and simultaneously a certain fascination with the sídliště that continues to this day – in his *samizdat* novel *Cesta českem našich otců* ([1983] 1992):

> You used to be able to take an intoxicating walk through the beautiful landscape and still be in the city. Now there are the sídliště ... Even the sídliště

> have their poetry. But it is a poetry of boredom or, at best, a poetry of pop-art. People go there to sleep, and other than a television they do not have much. Not a pub, not even a cinema. A smattering of green. The seamless blocks seem to stretch for kilometres, one after the other. The buses are packed beyond capacity … Somewhere tucked away are a few pre-war villas, and sometimes even a little garden pub. I'll sit there, the towering walls of the paneláky obscuring the horizon. The garden pubs are in bad shape, it does not seem to occur to anyone to take care of them … Surprisingly, their patios are half-empty, even though tens of thousands of new residents live around them. They are sitting at home watching television. Thirty years ago, there were many more of these garden pubs, and they were always full of people. Now, it is an exercise in melancholy just to sit there, but I don't mind it. I only hope that in all of those apartments people have at least two TVs. (pp. 20–1)

The crisis of the sídliště, as evoked by Bondy, did not concern the individual dwellings. Aesthetic criticisms aside, Musil notes that people on the whole were satisfied with their brand new mass-produced apartments and especially the many modern amenities they were not accustomed to having, like heat, hot water, and a private bathroom. Families also valued their newly won privacy, given that before receiving their apartments they were often sharing small apartments with their extended families.[1]

The problem, more specifically, was related to the space of the sídliště as a whole, particularly in the relation between dwelling and mobility, and the lack of communal and semi-public spaces. In his study, Musil suggests that the sheer quantity of housing does not alone define the post-war sídliště, which introduced completely new relationships between dwelling, shopping, and open space, marked most importantly by the disappearance of the street (1985, p. 15).

Hrůza places the crisis of the sídliště firmly within its social spaces: community spaces that traditionally serve both a social and a transiting function, like the street and the square, are missing from the sídliště (1967a, p. 1). Following a long line of modernists, he argues that in the building of the modern city, traditional streets, where different functions coexist, are "unworkable." Modern forms of traffic had made that mixing up untenable, so it was necessary to find a new solution. But in the new sídliště, where separation of pedestrian and car traffic had become the norm, all attention was given to the function of transportation – usually, car, bus, and metro – while the task of replacing the social function of the traditional street had, regrettably, been "forgotten" (Hrůza, 1967a, p. 1). Hrůza argues that the endless

Figure 4.5. The view from the highway entering Prague from the south, with South City looming on the horizon. Source: Steven Logan

pedestrian pathways through wide open spaces that permeate the sídliště, and many developments like it around the world, do not make up for the social space of the street because they are above all about getting from one place to another in the shortest time possible.

In *Architektura ČSR*, Miroslav Gottlieb and Marketa Todlová (1969) add their sociological perspective to the "interdisciplinary orchestra" addressing the building of new towns and sídliště. In a seeming critique of the megastructure ideal, the authors suggest that the sídliště need more than just one large complex where cultural and community amenities are located. Echoing Hrůza's critique, they argue that the sídliště also need lively pedestrian streets where people can sit outside at cafes and restaurants, and not just "corridors" that channel people from one destination to the next: "if our way from work was lined with cafes and restaurants, where it would be possible to sit outside in summer, if we met lit shop-windows and crowds of people in front of the cinema, we would have a different relationship to our dwellings and space in general than in the (best possible) case of differently painted facades, which should break the unending monotony" (Gottlieb & Todlová, 1969, p. 213).

The crisis of the sídliště, and the seemingly open way that architects and planners discussed the crisis, illustrates the environment of 1960s Czechoslovakia. Beginning in the summer of 1956, recalls architect Miroslav Masák (2006), there were increasing calls for "ideological and economical renewal and greater openness towards different views" (p. 11). During the Prague Spring, the Czechoslovak Union of Architects was one of a number of increasingly autonomous unions (which included those representing writers, musicians, painters, sculptors, and architects) attempting to reform socialism. In architecture and urbanism, the open atmosphere was reflected by increased international cooperation and a renewed optimism in the search for new ways of building socialist cities, whether it meant experimenting with new kinds of housing or attempting to enliven the public space.

The Plan

South City was built on the site of already existing independent communities, Chodov, Haje, and Litochleby in the south-east part of Prague (now part of the Prague 11 district). In 1922, Prague expanded its borders to include 13 districts and 19 city neighbourhoods, but according to local chronicler Jiří Barton, these villages did not want to join the agglomeration and so they remained outside the city's borders (1996, p. 67). However, the owner of a large part of the land in Chodov sold 294 hectares of fields to the city, which in 1930–1, used it to build 25 houses in what became known as the *kolonie*, the area's first suburban settlement. The city brought electricity and a bus line to Chodov in what was now becoming a suburban village with a municipal hall, church, theatre, and school (Bartoň, 2014). This building was part of a trend in Prague of building suburbs, usually with large family houses in the 1920s and '30s (Sýkora & Mulíček, 2014, p. 135). The areas around Chodov included the suburbs Zahradní město (Garden City) and Spořilov, both constructed in the 1930s with family houses. These neighbourhoods now contain both their old and new suburbs side by side. In 1968, Prague widened its borders for the first time since 1921, annexing 21 independent municipalities, including Chodov, Haje, and Litochleby, with the intention of using the land to build much-needed new housing (Bartoň, 1996, p. 75).

In 1965, the Department of the Chief Architect of Prague announced a design competition for South City.[2] After a two-stage competition that included 41 entries, the results were announced in 1967. The winning design (shown in figure 4.6) came from Prague architect Jan Krásný, a professor at the Technical University in Prague, and it was then passed on to Jiří Lasovský's atelier to make an urban plan.

Figure 4.6. The winning design in the South City competition. From its inception, the area with the highest density in the proposed city would lie within the bounds of the ring road. Source: Courtesy of V. Rothbauerová

In his review of the entries, Voženílek writes that the site of South City was chosen carefully so it could be an urban entity unto itself but also well connected to the city, thus avoiding the pitfalls of the existing *trabantni* sídliště, seemingly a direct reference to the Trabantenprinzip discussed above. The location was also chosen so that South City could have "direct contact with the open landscape" (Voženílek, 1967, p. 91): to the west of the planned zone of light industry was the Kunratický forest-park, to the south-east the Milíčovský forest, and to the north-east, the Hostivař recreational area, complete with an artificial lake (see figure 4.7), all within walking distance of the different neighbourhoods. South City would literally be a city in a park to which all residents would have access. The site was also chosen for its proximity to Czechoslovakia's first highway, the D1, running between Prague and Brno, which would divide South City's residential zone from the planned administrative and industry zone (work began on the highway in 1967). South City would become a "gateway" to the city on the Prague–Brno highway (Voženílek, 1967, p. 91). Modern architecture, it is said, is best viewed from the airplane, where the composition of buildings, roads, and empty space can best be appreciated. Brasília, of course, is a classic example.

Figure 4.7. View to South City from across the artificial lake in the Hostivař recreational area. Source: Steven Logan

Although South City's architecture can be taken in from the bird's-eye view, it was meant, rather, to be seen from the road on the approach to Prague (see figure 4.6). In this way, South City, as part of an explicitly new city-building project that rejected past urban forms, would form the entry point to socialist Prague (Voženílek, 1967, p. 91).

The entries to the competition for South City offered different ways to solve the many problems of the sídliště. In one design, residents would live in seven megastructures, each with 60 floors and 10,000 residents. The design was likely inspired by architect Karel Honzík's domurbia, which he developed in reaction to the growing distances between shops, cinemas, cafes, and other amenities: "These days, one has to go two, three or five kilometers just to buy some writing paper" (quoted in Hrůza, 1967b, p. 151). Honzík described domurbia as a complex of buildings that would contain as many urban amenities as possible "under one roof": administrative offices, apartments, artist studios, workshops, shops and services, and cultural and recreational spaces. The concentration of activities in one place frees up the surrounding land for either agriculture or recreation (Hrůza, 1967b). Domurbia was a variation on the megastructure, very much current at the time Honzík was writing, and yet these ideas were largely rejected by Voženílek, who noted that the entries to competition offered numerous different forms of dwelling that "questioned the idea of [a] large concentration of residents in domurbia." Although domurbia was rejected, the idea of the multilevel metropolis would play a central role in the envisioning of South City's centres.

Like Etarea, the conception for South City was in part motivated by new demands for personal mobility, but also the growing concern of people leaving the city for country houses. The simple projection in the official planning documentation of one car for every 3.5 people dictated much of the look of South City; in fact, the ratio exceeded the car-ownership rates of any country of the time aside from the United States and Canada, and was not reached in Prague until the early 1990s (Pucher, 1999, p. 227; 1990, p. 281). In 1960, Czechoslovakia had only 14 cars per 1,000 residents, whereas in France that number was 130, in Canada 224, and in the United States 345.

To a large degree, however, South City, like Etarea, was about reducing car use to a minimum and maximizing pedestrian movement within and among the suburb's communal spaces. South City's four neighbourhoods – Opatov, Litochleby, Haje, and Chodov – would be built so that each could accommodate around 20,000 people, a variation on the mikroraion. All through traffic in the residential areas would be pedestrian (see figure 4.8). An extensive pedestrian network would link South City's four neighbourhoods. Reflecting Ebenezer Howard's plan for his garden city (Fishman, 1977, p. 41), each district would have its own local centre, and one main centre would serve the whole development. The local centres would be situated above the main thoroughfares and, like in the plans for Etarea, would entail a complex network of above-ground, pedestrian-only walkways and shops. The main city centre, South City's downtown, would be located at the Opatov (formerly, "Friendship") metro station.

South City's city centre was also to be important as a secondary or subcentre to the Prague city centre. Its position next to the highway would allow it to be a "go-between" or "intermediary city," for people from surrounding villages to come and shop, and a marketplace for farmers from those villages to sell their produce (Gottlieb & Todlová, 1969, p. 214). With its city centre it could be a "centre of lesser importance for the suburban surroundings" (Gottlieb & Todlová, 1969, p. 214). Once it was established that the metro would be extended to South City, its planners also envisioned it as a place for people to park their cars and then head to Prague, or as they described it in a 1970 *Večerní Praha* interview, a place for people coming or leaving Prague to "have a break or even spend the night."

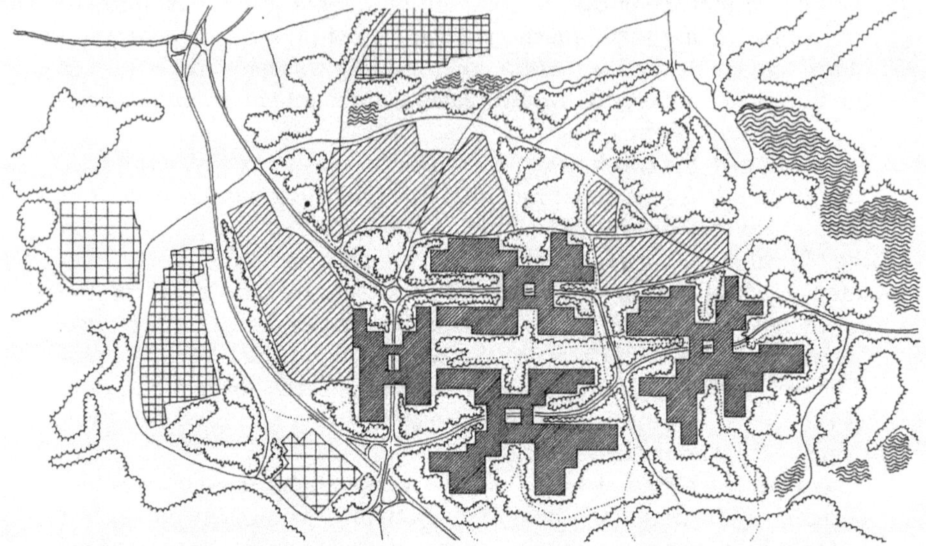

Figure 4.8. Schema for South City's urban plan. The darkly shaded areas show the four neighbourhoods of South City, and the dotted line shows the pedestrian network between and around the buildings. The other shapes show the settlements of the villages of Chodov, Haje, and Opatov, for the most part left intact. Source: Courtesy of Jiří Lasovský

In an attempt to create a diverse environment, the four neighbourhoods of South City would also be designed by a different collective of architects and would each use different building materials and technologies – prefabricated concrete panels as well as brick and poured concrete, for example.[3]

In keeping with the multilevel metropolises of the 1960s and '70s, South City was to offer "parallel cities" for cars and pedestrians, putting pedestrians above ground on an intricate network of walkways, while also attempting to construct on-the-ground pedestrian streets between the buildings. The plans for South City's main downtown and local centres represent a significant contribution to the heart-of-the-city discourse, but instead of being part of the redevelopment of a downtown, this was about the rethinking of peripheral space, the space of the sídliště.

Lasovský drew on his experience with places in the West. In 1965 and 1966, he visited the suburbs of Vällingby and Farsta in Stockholm and

Cumbernauld and other new towns in England. He learned from these trips of the new need for multifunctional urban centres that catered to pedestrians.[4] And much of Lasovský's writing around the time of South City's design and construction was critical of the earlier sídliště and the rigid application of CIAM principles. In response to the "deurbanizing tendency of the Prague sídliště with their atomizing division of functions," he looked for "another philosophical concept of the city" on which to base South City's designs (Lasovský, 1975, n.p.). There is an implicit critique here of the separations of the Athens Charter, which had already begun in the early post-war period within CIAM itself and in particular at CIAM 8 and with the work of Team 10. Lasovský claimed that "alienation, dehumanization, and loss of identity" associated with the earlier sídliště "would have no place" in South City (1973, p. 1). Lasovský's reference to alienation and dehumanization points to the larger discourse in Czechoslovakia on the search for socialism with a human face, but also the overall discourse in the East on architecture and humanism (Moravánszky et al., 2017). Lasovský has always emphasized that South City would not just be a place that people would inhabit (*bydlet*), but also where they would feel at home (*doma*).

The goal in South City was multifunctional centres in which people would work close to their dwellings and be near shops and cafes easily reachable on foot on what Lasovský called the *obytná ulice*, or living street.[5] The greenery from the surrounding forests was to extend right into the centres of the neighbourhoods, a feature that many Czech architects admired in Scandinavian post-war urban developments.

Although he did not subscribe to the rigid division of functions characteristic of Athens Charter urbanism, Lasovský still made a clear separation between movement within South City on the habitable streets and movement without on the major ring roads surrounding the neighbourhoods. One of the main goals of South City's planning was to minimize automobile movement within South City by offering parallel infrastructures for pedestrians, cars, and public transport. The key to the level of automobilization – one car for every 3.5 people – was the series of parking garages in the downtown, in each of the local centres and in the periphery around the four neighbourhoods, freeing up the areas between buildings for pedestrian use.

The neighbourhoods were composed in such a way that the tallest buildings were clustered around the local centres, and the further from the centre one got, the less dense and tall the buildings became until one reached the open spaces and the one- and two-story row housing on the periphery. The contrast in densities would answer Hrůza's call to address the uniformity of the sprawl of the paneláky on the existing sídliště. To bolster its labelling as a socialist suburb, South City was one of the few Prague sídliště where single-family houses were built; in the 1970s and '80s, single-family housing made up only 7% of new housing construction in Prague (Maier et al., 1998, p. 60). These houses were to have their own parking garages and would be built directly adjacent to the green spaces. Car and pedestrian traffic were to be mixed on the periphery. Smaller shops would dot the periphery so that residents could buy essential goods without having to go to the centre. In this way, the plan for South City brought together aspects of both the city and the traditional suburb.

According to official documentation, the pedestrianized living streets were to be a new way of approaching the Prague sídliště, becoming "the distinct element in the formation of the environment."[6] It was on these living streets, separated from car traffic, that Lasovský believed the possibility for spontaneous encounters would emerge, places to play chess, to sit and read, "places for people-watching." The living street, as Lasovský understood it, was similar to the way Jaqueline Tyrwhitt understood the core, and as such it evokes an idea and a planning principle that would continue to be important in the sídliště planning in the 1970s and '80s. Tyrwhitt affectionately describes one iteration of the core as "the daily or weekly social promenade" where people "turn out to greet their acquaintances, observe their neighbours, gossip with their friends, and meet their sweethearts" (1952, p. 103). The living streets would be complemented by a system of recreational pedestrian paths that would follow strips of greenery that extended from the peripheral landscape, allowing pedestrians to walk around the city or to the park and out into the forests "without conflict with traffic and from most parts without any interruptions in the path" (Lasovský, 1975, n.p.).

In this sense, the separation of traffic, a feature of so many modernist plans derived from the Athens Charter, also attempted to foster new kinds of communal spaces outside of the traditional streets of the older downtowns and historic centres. In their description of the detailed urban plan, Lasovský, along with fellow architects Jan Krásný and Miroslav Řihošek, echoed planners going back to the Radburn plan:

> the most difficult task ahead is minimizing the harmful effects such as noise, vibration, exhaust, and traffic collisions on the residential environment. The most promising way to restrict such effects is the complete segregation of pedestrian and transportation spaces. There is a purity to the zoning plan in terms of pedestrian traffic and it is the result of both sociological and architectural work. The pedestrian paths and spaces create a self-contained system. (Lasovský et al., 1969, p. 446)

Although automobile use was to be minimized within South City, like other cities – both socialist and capitalist – the design was dictated by what was seen as the inevitable future growth in automobile use. The authors of a 1970 article on South City in the newspaper *Lidová demokracie* ask: "Does the emphasis on pedestrian traffic mean that in the time of the explosion of automobilization residents will be doomed to just walking? Never." Here, the planners rework Clarence Stein's dictum that architects think "not how to live with the automobile, but how to live in spite of it."

While it was being planned, South City was frequently compared to Olomouc, a medieval city with a similar population. The comparison between Olomouc and South City was the basis for the challenge posed to the architects and planners, put in the following way by Lasovský: "build in 15 years what usually takes 800."[7] In a 1971 PPÚ publication, the architect and theorist Otakar Nový explicitly evokes the difficulties of reconciling the art of city building with industrial production: "Great works, which our ancestors needed hundreds of years to make, we are today capable of managing in a few decades thanks to socialist planning, typification and industrial methods" (n.p.). Nový goes on to ask (perhaps rhetorically): "Is it really possible to compare great works thousands of years in the making with a building project of three, four, or five decades?" As Lasovský's statement implies, the only way to realize the vision of a city in such a short time is to use the materials and techniques of modern industrial production, and yet it was precisely these new technologies that gave architects the opportunity to design entirely new cities on such a large scale. Although Expo 67 offered a utopian take on the instant city built using new methods of production, South City presents a much different story, one in which construction cranes, reinforced concrete, and prefabricated concrete panels, as well as the technical knowledge for building, transporting, and assembling them, took priority over the visions of the architects and their plans to reform the sídliště.

Implementation and Critique

"Only visions." This is how Vítězslava Rothbauerová, a participant in the design competition and the chief architect of a later additional development in South City in the 1980s, South City II-West, described the many ideas for South City.[8] The post-war demand for housing shortage in Prague provided a significant opportunity for architects and urbanists to design entirely new cities. With a planned socialist economy, and the state holding all the land, this process was made even easier and more attractive, but that opportunity came about precisely because of the pressure to meet the state housing quotas for production. Rothbauerová's own experience is illuminating:

> Together with the suppliers and investors, we agreed to develop a new kind of panel building for this [a second South City] development. We were excited. Of course, this promise was quickly broken as the construction workers already knew by heart how to assemble these buildings and they were not going to burden themselves by having to actually look at the drawings. There were five types of the VVU-eta system for assembling apartments and they were repeated throughout the whole country. (Rothbauerová, personal communication, 8 June 2012)

The actual building of South City followed the Soviet occupation of Czechoslovakia in 1968 and coincided with the wider purge of the reformist elements from the 1960s that took place in 1971. As Hrůza recalls, the Office of the Chief Architect was very much affected by the Soviet occupation – Jiří Voženílek, although close to retirement age, was forced to resign his post as Prague's chief architect and was no longer allowed to lecture or teach (Hrůza, 2006, p. 37). After the occupation and the political purges, which included changes in the leadership of the PPÚ, Lasovský was forced off the project as chief architect in the spring of 1970. His atelier was also disbanded (its members had been printing anti-occupation material in their offices).[9]

When construction began in South City in 1972, the state construction company had ceased to "coordinate on the ground" with either the architects or the PPÚ (Lasovský, 1975, n.p.). South City would set a frantic building pace. In a panel discussion in the journal *Československý architekt*, Lasovský claimed that "the Prague Fathers want to set a record in South City. As opposed to the roughly 2000 apartments delivered in 1969, they want to build 5000 apartments yearly" ("Jižní město Pražské"). Although Lasovský and the other architects had imagined different building methods and materials for each of the neighbourhoods, only the prefabricated concrete panels were used, and the only difference was length – the ideal was the infinitely long building with almost no variation from one section to the next – and height: 4, 6, 8, or 12 stories. Lasovský points out that the unbelievably long buildings that characterize the sídliště were often a result of economizing on costs in what was an otherwise very expensive and complex process: it was cheaper to keep building, or to build in the immediate vicinity of the crane, than to have to move the track upon which the crane moved. And the crane works best when the tracks are straight, so the ideal building in a city of crane urbanism was the endlessly long apartment building.[10] These were not towers in the sky; they resembled a giant wall more than a tower.

In order to meet the demands to build so many buildings annually, "every two minutes a 10-ton truck would enter South City with a truck-full of gravel" ("Jižní město Pražské"). Later in his life, Jiří Voženílek admitted that the biggest mistake of his long career was his support for industrial house-building technologies because they had to be pre-fabricated in large factories and transported significant distances to the construction site; the heaviness of the panels on the trucks, he would come to learn, destroyed the unprepared road system. It was not industrialization of dwellings per se that was the problem, but their effect on the surroundings.[11]

This process is captured most profoundly in Věra Chytilová's film *Panelstory, or How a Sídliště Was Born*, shot during South City's construction in the late 1970s.[12] South City's lunar-like landscape is not simply the stage on which the film unfolds. Although the film includes a cast of characters – the people living in the sídliště and trying to negotiate the muddy landscape and the workers building it – the actors are also the cranes, exposed pipes, mounds of construction material, panels flying through the air, and piles of excavated earth surrounding the apartment buildings. What marks South City's destruction is not that the former villages were levelled to make way for the apartment blocks – some homes were lost, but many of the original houses of the villages remain intact – but that people moved into the sídliště before it was finished; mothers negotiate makeshift sidewalks with baby carriages and confused people try to locate apartments in a landscape in which there are neither street signs nor apartment numbers.

Destruction here works on two levels: one, in the sense of liquidating the traditional city streets in favour of a new kind of city and new kinds of pedestrian and public spaces, and two, the actual destruction of the landscape to make way for the roads, the subway, and apartment blocks. Although Lasovský envisioned the separation of cars and pedestrians, he also imagined within those pedestrian spaces the intertwining of green spaces and dwelling spaces, with the greenery on the periphery permeating to the doorsteps of the buildings. This is what gave so many of the sídliště their characteristic barren look in their early years and what distinguished them from their Scandinavian counterparts.

In an article criticizing South City's realization, Lasovský wrote that the hallmark of the plan, the living streets with preference given to pedestrians, "disappeared," and the overall composition of the plan – that is, the contrast between the densely built-up centres and the open spaces of the periphery – was lost because of the "considerable pressure of the building technology, deforming the attempts at city building." He wrote that the "final impression" was one of an "inhospitable, uninhabitable" environment (Lasovský, 1975, n.p.).

Although it was hoped that these living streets would be lined with small shops and cafes – the urban parter that Lasovský believed was so important for addressing the crisis of the sídliště – in the period of normalization, recalled Rothbauerová, there was "little street life, and so it was easier to have everything in one big building." The thoroughfares that run through and around South City are just that; as the planners had anticipated, these were not meant for pedestrians, and so there is little chance for convivial encounters or spaces that one would want to linger in. All transport was put underground rather than on trams and light rail, as some of the original designs had proposed. The living streets became clogged with parked Trabants, Skodas, and other cars endemic to the Eastern Bloc as the above-ground parking garages never did get built and the surface parking lots on the periphery of the neighbourhoods were deemed unsafe – for the cars, that is – and too far from the apartments (see figure 4.9).[13]

Although socialism emphasized both the quality and quantity of the living environment, the public over the private, crane urbanism troubles these easy distinctions. Crane urbanism did not disable the building of the massive green spaces typical of the socialist suburbs; in many ways, the scale of crane urbanism matched the scale of these huge open spaces (see figure 4.12). And although, as Hirt notes, wide open spaces were typical for modernist developments in the West as well, it was the sheer extreme of the socialist divide between public and private space that stands out, and in particular, the lack of "in-between" public spaces like cafes, small shops, or balconies. (2012, p. 40). These in-between spaces were most lacking in South City, along with a dearth of non-essential collective facilities that were to define the socialist living environment.

David Harvey singles out the image of creative destruction as key to the "practical dilemmas that faced the implementation of the modernist project" (1989, p. 16). In South City's case, the dilemma, posed by Lasovský, was how to build a city in 15 years when that would usually take 800. The figures Harvey chooses as emissaries of this creative destruction – drawing on Berman (1983) – are now ubiquitous in the history of modernist urbanism: Baron Haussmann, Le Corbusier, and Robert Moses. The artists, architects, poets, and philosophers play heroic roles in envisioning the modern future, although Harvey

Figure 4.9. Typical "street" in today's South City. Source: Steven Logan

Figure 4.10. Neglected above-ground pedestrian space. Source: Steven Logan

Figure 4.11. Incomplete walkway system. The bridge in the foreground was to connect to the bridge in the background. Source: Steven Logan

Figure 4.12. South City's Central Park. Source: Steven Logan

notes that the results may be "tragic" (1989, pp. 18–19). The models for Benjamin's "destructive character" were Loos, Klee, and Le Corbusier, those who were not afraid to reject the past and "start from scratch" (Benjamin, [1933] 1999, p. 732). Lefebvre notes that dominant space is "invariably the realization of a master's project" ([1974] 1991, p. 165), but his reference points were, like Benjamin's and Harvey's, the heroic modernists of the interwar period.

Harvey's creative destruction and Benjamin's destructive character are defined by their creative visionaries, but what if the agent of creative destruction is not simply the heroic architect of modernism, but the very tools and technologies through which the modernist project is realized (and without which this heroic architect is rendered impotent)? If the main figures, and scapegoats, of architectural and urban modernism are so often its male anti-heroes, who do we turn to in the case of South City? The lack of heroic architecture was part of the point of the socialist turn away from individual star architects and towards collectively authored projects – the prominent role of the crane and machine technologies is one outcome of this anti-heroic architecture.

Although the general layout of South City accords with Lasovský's original master plan, most, if not all, of the architectural details were ignored. To hear the architects' side of the story, one could plausibly claim that the crane had taken Lasovský's job as chief architect. Peter Lizon, an architect who took part in the South City competition, writes: "the creative freedom of the site planning design was strictly dictated by the runs [tracks] of the construction cranes lifting the heavy concrete wall panels" (1996, p. 109). Lasovský claimed that there was a "total absence of architecture" in South City.[14]

In her history of housing in Czechoslovakia from 1945 to 1960, Zarecor points to the tension between the architecture of the unique building and an architecture of manufacturing, mass production, and materials. She writes that when the architect ceases to practise his or her art, then architecture becomes "mere 'building' instead of architecture with a capital A" and is written off by historians as simply "economic and technological determinism" (Zarecor, 2011, p. 296). It is no wonder, then, that architects would resort to such strong language – "strictly dictated" and "determined" – when describing crane urbanism, since for them it is simply the state and the crane as the arm of the state imposing its will. The main actors on the scene of crane urbanism, aside from the cranes, were the workers who committed to memory the Lego-like assembly of the prefabricated apartments, the unnamed inventors and engineers who came up with the particular panel technology used in South City, members of the state building company that pressured the architects, and, of course, the machines.

From Lasovský's perspective, there is no "architecture" not simply because the panel technology was being used – here it is useful to return to Lasovský's advocacy for the urban parter, the life between buildings. For Lasovský this was part of the domain of architecture, the living environment, without which the socialist city is incomplete. In his history of concrete, Adrian Forty (2012, p. 249) notes that the biggest threat to architects in the post-war period was the system building of the apartment blocks, where the architect became simply a "technician," whose main role was the site arrangement of the buildings.

The architects were not against prefabrication per se, as the remark from Rothbauerová quoted above suggests, and they were not only technicians. In the tradition of socialist city building, they were to compose the city as a work of art. The death of the street as proclaimed by modernists of both the socialist and capitalist persuasion had its corollary in the "the organic unity of the architectural ensemble" (Maxim, 2011, n.p.). At the scale of the ensemble – the mikroraion, the sídliště, the city centre – the architects could transcend their roles as technicians and facilitators of mass production and "reach into the 'ideological and artistic realm' through compositions at the city scale." (Maxim, 2011, n.p.). This is certainly how South City's centre can be perceived and appreciated. The challenge was to use new techniques of production (quantity) in combination with the living environment suitable to late socialism, to bring together aesthetics, ideology, and industrialization (Krivý, 2016, p. 96). These were in dialectical tension in South City: the city as work of art in the age of mass production.

The City Centre

Although he was forced to resign his post as chief architect, Lasovský continued to be involved with the project to design and plan subsequent iterations of the city centre up until 1981, when the last designs were made. The city centre was to be the main element in the total work of art of South City, worthy of CIAM's plea to architects and urban planners to attend to the heart of the city, and also devised in an atmosphere of interdisciplinary collaboration. Together with scenographer and national artist Josef Svoboda, who himself played a key role in the art direction of the very popular Czech multimedia installations at Expo 67, Lasovský came up with a plan for the city centre. Svoboda and Lasovský elicited suggestions from artists, sociologists, and psychologists

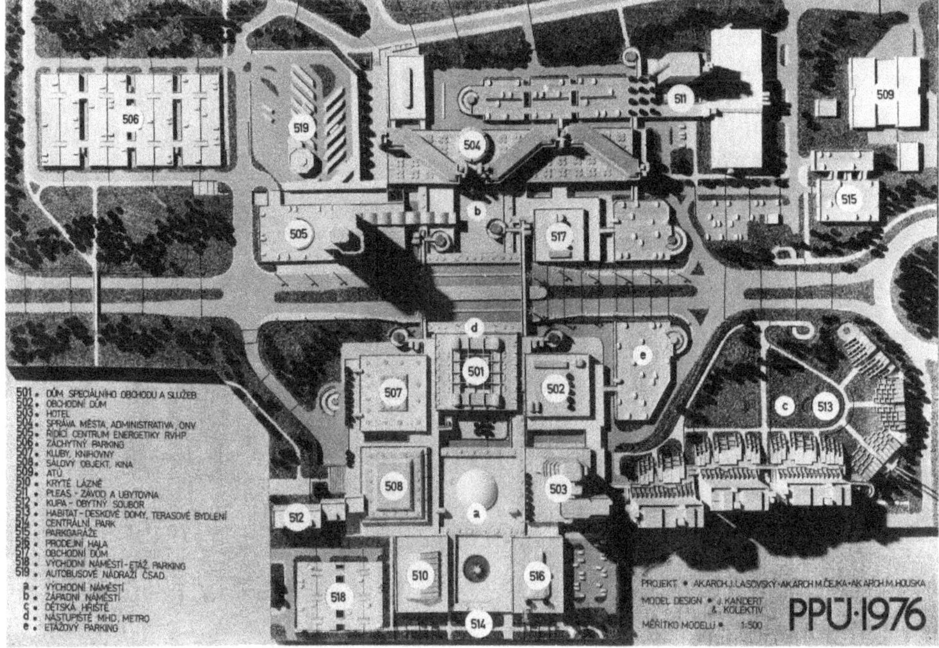

Figure 4.13. Lasovský's plan for South City's city centre, 1978. Source: Courtesy of Jiří Lasovský

when they were planning the centre. This interdisciplinary approach, although influenced by the interwar avant-garde, achieved particular prominence at Expo 58 in Brussels. The architects of the popular Czechoslovak pavilion sought out a new style that reflected both the Bauhaus interest in interdisciplinary cooperation and Khrushchev's call for works that embodied the collective spirit of socialism rather than the desires and whims of individual architects. The architects František Cubr, Josef Hrubý, and Zdeněk Pokorný wanted to achieve a "complex work, resulting from the widest cooperation with painters, sculptures and graphic artists" (quoted in Giustino, 2012, p. 195). This collectivism would distinguish socialist architecture from the star architecture of the capitalist system and make the building, not necessarily a work of art in its own right, but part of the larger project of building socialism through architecture.

Lasovský's plans for the city centre were featured in an article in *Československý architekt* called "Centre for 100,000" (Novotný, 1978). The most prominent feature of the proposed centre (shown in figure 4.13), according to the article's author, was that it negated "all the laws of functional zoning" combining shopping centres, cinemas, cultural centres, and a public square with a retractable domed roof, a factory, "a hot-dog stand next to a jewelry shop," administrative buildings, and a residential area with low- and high-rise buildings, entitled "Habitat" (Novotný, 1978, p. 5). The entire centre was to be reserved for pedestrians, "without a single conflict with other transportation modes" and a variety of pedestrian routes to choose from. South City's centre would be a "transportation paradise."

Anchoring the city centre and the Friendship (today Opatov) metro station to the west and the Kosmonaut (today Haje) station to the east would be Central Park: 1 kilometre long and 100 metres wide, built with the earth that had been excavated to build the metro and was accumulating in the space where the park was to be. For Lasovský, who believed it would be a waste of energy to truck all the soil away, the "dirt was the inspiration" to transform the lunar landscape into a "supersculptural park" that would be a "symbol of the Czech landscape, with its hills and valleys, footpaths and passageways" (Novotný, 1978, p. 5). South City's city centre and Central Park would continue the socialist practice of building green cities while also developing the city-centre models of the heart of the city. The city centre, the park, and its surrounding apartment blocks would complete the physical transformation of the sídliště into a socialist city. The plan for the city centre was a nod to Expo 67, with its Habitat residence, but also a reflection of many of the utopian ideas of the 1960s. The author of the *Československý architekt* article plays up the contrast between the sídliště as strict product of industrial technologies and the vision of a new city: "Ten years ago, when the project originated … it imagined building a city," and yet the completed areas "are still reminiscent of a sídliště" (Novotný, 1978, p. 4).

Although Lasovský was given the task of designing and planning the city centre and South City's Central Park, in the absurdist fashion that befit the normalization period, he knew it was never going to get built because there was only money for building the dwellings and the most important facilities. In spite of that – or because of it – they were *really elaborate* plans. Throughout the 1970s, Lasovský and his team continually developed the plans for the city centre, as the studies were funded by grants from departments like the Výzkumný ústav výstavby a architektury, which was founded by Voženílek in 1951 and at which Jiří Musil worked at the time. As the industrial logic of crane urbanism became ever more *obscene* and destructive in the 1970s, Lasovský, along with the artists he brought

Figure 4.14. The in-between spaces in South City II-West. Source: Courtesy of V. Rothbauerová

on to work with him, in ever more fantastical ways attempted to construct the ideal urban *scene*, a kind of phantasmagorical counterpart to what Lasovský saw as the damage inflicted by South City's crane urbanism.

At the same time that Lasovský was at work on designs for South City's city centre, his fellow architect Vítězslava Rothbauerová was working on a second South City development, situated to the west of the D1 highway on what was supposed to be the area devoted to light-industry jobs for South City's residents. In many ways, this second South City development, known as South City II-West and built in the 1980s, sought to respond to the deficiencies of the sídliště and is an important development in the late-beautiful phase discussed in the previous chapter. Rothbauerová and her team, as suggested above, sought to work within the confines of industrialization rather than against it. The development included coloured facades, an orientation system in which each neighbourhood in the sídliště would have its own symbol, and a greater attention to the parter, the space between the buildings (see figure 4.14).

Conclusion

South City, in its design, implementation, and lived reality, is an exceptional example of the socialist suburb, and although peripheral to suburban histories, it is central to global suburbanization in the 1960s. It drew on a wealth of material from the West, and not simply the Athens Charter urbanism of early CIAM; rather, the designs reflected the humanizing gestures of CIAM's post-war architects, as well as the heart of the city discourse, that was important in both the East and the West, whether in new towns like Cumbernauld or suburbs like Vällingby. And at the same time, in terms of prefabrication, South City was to be one of largest and fastest-built projects, and so it became an emblem of the new socialist city of the post-war period, marking the move from mere sídliště to socialist city. After all, its position next to country's main highway did not simply serve a functional purpose: anyone driving into the city would be welcomed by this new gateway to the socialist city.

And as Hirt and others argue, the attention to building dwellings and essential infrastructure like clinics, food shops, etc., meant that people often moved in to an unfinished work of art. This was especially the case with South City, as the infrastructure costs alone were immense. In particular, building the metro out to South City presented quite a challenge: the further from the city, the more expensive the new sídliště became, which demanded large, flat areas of green spaces that could only be found on the periphery. The metro had to be extended 5 kilometres out to South City, and although the state claimed it was the most inexpensive option, "this was an illusion," recalled Hrůza. Like its capitalist counterparts, it was an economy based on "extensive growth" (Hrůza, 2006, p. 37). Importantly, however, this was growth not based on the automobile. Although some of the initial designs submitted for the competition planned for a form of light rail with multiple stops within the sídliště, the subway today is one of the most attractive features for South City residents, the most important infrastructure linking it with the city centre.

Although South City, as its name implies, was supposed to be a city, and there were initially plans to make it a satellite city, it was never to be its own town; with the extension of the metro to South City in late 1980, its dependence upon and connection to the city was complete. South City's very name, and the way it was understood after it was built, implies incompleteness, the suburban as somehow less than urban – it is a reference to an elsewhere that South City needs in order to exist. (An alternative translation of the Czech rendering of South City is Southern City.) In the initial competition, South City was to be a self-contained satellite city that at the same time would offer an easy connection to the city. Very early on, South City was defined by its architects, and increasingly its inhabitants, in relation to an elsewhere, to either the centre of Prague or the landscape of chaty beyond.

South City's reference to an elsewhere (and necessarily an "elsewhen") highlights its importance to the alternative histories of both modernist urbanism and the suburb. Depending on who you talk to, South City owes its existence to another place, be it Etarea, Milton Keynes, Vällingby, or Cumbernauld. South City was formulated, particularly in the 1960s, according to Western planning principles as much as it was a model of mass production and a model socialist city. And at the same time, with the relatively low penetration of the automobile in socialist countries, South City had an opportunity to become a pedestrian city, where, as Lasovský had hoped, car use would be reduced to an absolute minimum.

On the one hand, South City could be thought of in the classical sense of a suburb with apartments in place of single-family houses. In her work on chata culture in the post-1968 period, Paulina Bren writes that a "good communist" under normalization "defined himself within the contours of his private life" rather than public life as it played out on the city streets (2002, pp. 126–7). Gustav Husák, as he assumed control of the Communist Party in the wake of the Soviet occupation, guaranteed that the party would "safeguard the quiet life" as a response to the events of the Prague Spring – as he put it, a "normal person wants to live quietly" (quoted in Bren, 2002, p. 123). The quiet city favours the private life of the apartment, offering people refuge from the cold order of normalization, and the subway an easy way to get out of the city. Lasovský's critique of the sídliště and his attempt to create a place that people do not just inhabit but actually call home, also echoed the normalization policy: in 1982, Communist Party chairman Miloš Jakeš exhorted architects "to create a living environment conducive to happy family life ... where people would feel at home" (quoted in Krivý, 2017, p. 316).

As Slavoj Žižek has quipped, the last thing the Communist government of the 1970s wanted was for their citizens to be communists. This is not far from William Levitt's claim that no Levittowner who owns his own house could be a communist because he has too much to do: fix up his house, tend to his garden. As Bren argues, "the political ideology that became the cornerstone of 'normalization' was ... to define and locate communist citizenship within a publicly shared private world" (2002, p. 123). The domestic comfort and mass-produced "necessities" of suburban living so commonly associated with post-war suburbs like

Levittown had their corollary in the focus on domestic comforts in the socialist apartment. This was particularly evident during the consumer socialism of the 1970s, when magazines like *Domov* addressed "housing art and culture in the household." As in the post-war appropriations of the garden city, which neglected communal aspects in favour of universal single-family homeownership, the private life prevailed. The "romantic suburb," as Mumford (1938) understood it, was "a collective attempt to live a private life."

At the same time, South City and other sídliště like it offered an alternative to the dominant suburban narrative of single-family homeownership. The collective and communal aspects, although in part unrealized, along with the connection to the public transport networks and the rich green spaces that surround South City, suggest that the architectural criticisms of the socialist living environment and the absence of "living streets" is an illustration of the connection to the West as much as it is a critique of the sídliště's inability to become a city in the classical Western sense (even if the goal of the socialist city was to negate the capitalist city, not emulate it). South City, then, exists on the boundary of capitalism and socialism.

There is a way in which the privatized life of normalization in the 1970s can be seen as suburban, even if Lasovský's plan meant emulating city spaces. If the "living streets" never did materialize, as many of the commentators of the 1960s hoped for, the wide open spaces between buildings would define the public spaces of the suburbs. Although they are derided by critics of modernist urbanism, many of the respondents in Musil's study, conducted in the early 1980s, appreciated the spaces between buildings – indeed, they wished that the spaces would be even larger – and although the network of pedestrian pathways was not built according to Lasovský's plan, people still appreciate the separation between cars and pedestrians. There is an excellent public transportation network, which allows people to quickly and easily commute to the centre, as well as access to the countryside beyond either by bus or by connecting to nearby train stations. The freedom to live without a car and live with public transportation is a lasting legacy of socialist suburbanism, even if the nature of socialist planning meant that it might take a while for the amenities to come.

The building of South City's centre reflected trends then influencing the building of regional subcentres in both East and West. The desires of South City's architects and the planners reflected a larger modernist trend of constructing city centres in the suburbs, one of the key ideas of Canadian urban theorist Humphrey Carver, as we will see in the following chapter when we look at his own attempts to re-envision the post-war North American suburb.

5 Redesigning the Post-war Suburban Landscape

And always in the background there has been the unconsidered residuum of land, the confused and unplanned areas where the ordinary people have lived. For the interests of the common people have never yet dominated the urban scene. The contrasts of power and poverty, of wealth and humility, have commonly been regarded as inevitable features of our civilisation ... But it has been the instinctive hope of escaping from such a scene that has led to the peopling of this continent. The inspiration which led so many of our forefathers to come to Canada was the vision of a country in which the important elements of design were neither mansions and monuments of an upper class nor the factory chimneys and skyscrapers of big business. They came to look for a place in which the homes of the ordinary people would be the dominating element in the plans of cities.

– Humphrey Carver, 1939

While Prague architects and urban planners were criticizing the sídliště on the city's periphery, Humphrey Carver was offering his own critique of the Canadian suburban environment. The dominant element of Toronto's early post-war suburban landscape was the mass-produced single-family homes, the purchase, financing, and design of which was supported by the Central Mortgage and Housing Corporation (CMHC), a Crown corporation founded in 1946 and for which Carver worked from 1948 to 1967.[1] CMHC created a system of mortgage finance, which along with the rationalization of the building industry and new ways of subdividing land and building houses, helped create the new post-war suburbs (Harris, 2004); they are the Canadian suburban counterparts to the socialist concrete apartment blocks and brought a similar set of concerns among architects and planners. CMHC was central to the formation of the post-war suburbs, the legacy of which is still a part of the fabric of Canadian suburbs (Keil et al., 2015, p. 90; Harris, 2004).

Humphrey Carver has been described as one of Canada's "leading planning theoreticians" (Sewell, 1993, p. 44), the "most influential planner" of the early post-war years (Harris, 2004, p. 15), and "one of Canada's most respected authorities on housing reform and the building of suburbs" (McCann, 1999, p. 129). Carver was one of many British emigres who came on the Toronto planning scene in the 1930s. In his discussion of the diffusion of planning ideas in Canada, planning historian Stephen V. Ward (1999) argues that "CMHC set an important precedent by actively recruiting European, overwhelmingly British, planners. Personifying this new mood within the CMHC was Humphrey Carver" (pp. 64–5). Carver's work in Toronto during the 1930s, and with the CMHC in the post-war period, is central to the modern planning and theorization of suburban neighbourhoods.

The aim of this chapter is to use Carver's life and work to look at post-war attempts to foster an alternative to the classic sprawling suburb of single-family houses. Carver's work on social housing in the 1930s, and his turn to the suburbs in the post-war period with CMHC, reflects many of the dominant tendencies in the production of suburban space going back to the key moments of the 1920s. Although this chapter takes Carver's life and work as its focus, it is primarily about the suburban environment he helped create and then later criticize. Carver's often fraught relationship with the suburb is drawn out in this chapter by looking at his devotion to the family and the single-family house, and his interest in moving beyond Perry's neighbourhood unit as a structuring principle for suburbs.

In a speech to the Town Planning Institute, Stewart Bates, president of CMHC from 1955 to 1964, remarked: "In parts you are bureaucrats, committed to the daily task of moulding things into standard and uniform patterns. But in part you are designers and humanists trying to release people from the tedium of the mass-produced city" (quoted in Carver, 1994, pp. 55–6). There is a fundamental tension in Carver and his work at CMHC that will play out across this chapter: between the design, financing, and building of mass-produced houses and the attempts to foster collectivity and community through urban form. Carver offered that alternative in his visions for cities in the suburbs, centres for collective life and attachment.

The Interwar Period

Like previous chapters, this one begins with the 1920s, when Carver was attending the Architectural Association School of Architecture (AA) in London, and continues through to the 1960s and '70s, following Carver's travels from England to Canada and his career with and subsequent retirement from CMHC. While Teige was developing relationships with Bauhaus director Hannes Meyer and CIAM secretary Sigfried Giedion, Carver admired the work of the Bauhaus from afar, first in London and then later while working in Canada. Carver first came across the work of Le Corbusier in a Paris bookshop while he was on break from his second year of studies in London in 1926. He describes his reading of *Vers une architecture* as an "electrifying intellectual experience which immediately changed my whole way of looking at the world around me, at buildings old and new" (Carver, 1975a, p. 21). There is no shortage of lyrical praise for Le Corbusier and also the Bauhaus in Carver's writing. Counting Mies van der Rohe, Walter Gropius, and László Moholy-Nagy among his influences, Carver recalled in a 1968 speech that "we came to believe that the most exquisite beauty and social refinement is the polished product of technology, anonymous and shining, steel and glass."[2] In "Planning Canadian Towns," an unpublished manuscript abandoned at the onset of war in 1939, he elaborates on this change: "until Le Corbusier arrived with his dazzling and elusive logic, we did not know how to transmute the industrial city into a noble and poetic form, without losing its contemporary quality" (Carver, 1939, pp. 48, 50).

In this way, Carver took up his own position via the debate that Le Corbusier and Teige had begun back in the 1920s, with Carver attempting to negotiate the tension between Le Corbusier's view of architecture as the timeless and monumental art of composition and Teige's strict interpretation of architecture as instrument. In this aspect of the debate, Carver clearly sided with Le Corbusier, who emphasized that architecture was above all about composing the elements of the landscape. Le Corbusier argues that composition, which Teige dismissed as the "Godly mission of architecture" (Teige, [1929] 1974, p. 90), is actually the key to the architectural plan and the way things in space come to be "architectured" (Le Corbusier, [1933] 1974, p. 96). Carver believed that it was the architect's or planner's job to bring a sense of aesthetics and beauty to the anonymous products of industrial civilization: apartments, highways, office buildings. The contrasts between instruments and monuments, between the apartment block and the single-family house, between CMHC's "mortgage-instrument" and the art of urban planning, are central to Carver's approach to the post-war suburb.

After finishing school in London, Carver immigrated to Canada in 1930 and took up work with the landscape architect Carl Borgstrom (Carver, 1975a, p. 28). Carver considered himself a socialist and in his early work in Toronto in the 1930s, he became a housing advocate and was involved in the activities of the League for Social Reconstruction

(LSR), a group of left-wing intellectuals and academics formed in 1932 (and disbanded in 1942); they helped form Canada's main socialist party, the Co-operative Commonwealth Federation (now, the New Democratic Party, or NDP). Carver, and the people involved with the LSR, might best be described as "Red Tories" – that is, a unique English Canadian combination of conservative and socialist principles, the latter of which is "British, non-Marxist, and worldly" (Horowitz, 1966, p. 159). The Red Tory, writes Gad Horowitz, is "a conscious ideological socialist" who does not make reference to class struggle and firmly supports state intervention coupled with a "paternalistic concern for the 'condition of the people '" (p. 157). The Red Tory, Horowitz continues "is a philosopher who combines elements of socialism and Toryism so thoroughly in a single integrated *Weltanschauung* that it is impossible to say that he is a proponent of either one as against the other" (pp. 158–9). Carver's ideology combined a conservative belief in the family and the single-family house as the moral bedrock of society and socialist-leaning beliefs that the state has a responsibility to reform the environment of poverty in which the poor live. Carver did not question the economic system that produced inequality; rather, he sought, through architecture, to address these inequalities: by designing above-standard housing, he believed, social inequalities would be dissolved.

Carver got involved with the LSR through his brother-in-law, King Gordon, one of its founders. Carver contributed to *Canadian Forum*, the voice of the LSR in the 1930s, and to one of the LSR's main publications, *Social Planning for Canada* ([1935] 1975), in which he set out the "general principles of town-planning" and the need for a nationwide housing program while criticizing the suburban speculator and the slumlords of the inner city (Carver, 1975a, p. 51). Through his work with the LSR he secured a teaching post at the University of Toronto, where he took on a central role in addressing social housing in Toronto in the 1930s.

In a utopian fashion that befits the modernism of the interwar period, Carver called for "nothing less than the gradual reconstruction of the entire fabric of our civilisation." In the depression years of the 1930s, Carver clung to his "visions and utopias rather desperately" in the face of economic hardship (1975a, p. 46). In his office at the Housing Centre at the University of Toronto he covered the walls with images of public housing from the United States, as well as from London, Liverpool, and Vienna (Carver, 1975a, p. 51). When Carver proclaimed in "Planning Canadian Towns" that the goal of planning was to rebuild "obsolete" villages, to control and manage growth, and to "construct new settlements in which will be embodied the experience of the ages and the

Figure 5.1. A Toronto suburban house from the interwar period. Source: Steven Logan

hope of the future" (1939, p. 6), he was aligning himself with the modernists who believed the only way to accommodate the new technologies of transportation and contain the growing forces of urbanization was to rebuild the city anew.

Carver used the metaphor of surgery to explain the radical change needed, drawing on a history of modernist urbanism's use of bodily metaphors to describe ailing cities and clogged circulation.[3] He praised the cities that planned a "vigorous attack upon their obsolete central areas – to cut away the deadwood where decay has set in. City surgery is not unlike tree surgery. You have to prune away the part that has lost its vitality in order to get new fresh growth and bloom." Carver called for urban renewal in the so-called slums of Toronto: he believed that "city surgery" was required for the entire area south of Bloor Street, between Bay Street and the Don River (1948, p. 34). The culmination of Carver's work in the interwar period was his involvement in the renewal of part of this area, which eventually became Regent Park, a project of which Carver was particularly proud.[4]

House Lust

Carver entered the AA with a mission: to learn to "make beautiful houses for people to live in" (Carver, 1994, p. 23). Throughout Carver's work one can read the ways in which ownership of a single-family house on a plot of land was so central to the Canadian settler experience. Carver went so far as to name his fascination: house lust. The term never actually appears in any of his published work, but it does appear in his notes for a "CMHC Senior Staff Course" in 1957 and again in his notes for an aborted book project from 1965 that he tentatively called "A Pride of Cities." He defines house lust as "the enjoyment of a beautiful house [which] is *par excellence* a satisfying and intellectual accomplishment." House lust is a "deep and primitive urge" to "possess and beautify a place you can love." House lust brought Europeans to North America, according to Carver, since it it the "most important element in the life-style of any community" (Carver, 1957). In the notes for the 1965 book project, Carver identifies house lust, along with "social expression," the "conservation of resources," and the "City Beautiful," as making up CMHC's "evangelism" (Carver, 1965).

Behind this house lust was also a deep moral commitment to the house as the best place to raise a family. Opening the section on housing in "Planning Canadian Towns," Carver put forward the following proposition: "the Family is the biological institution around which our Housing must be designed and that, in order to express its corporate personality, each family needs an individual home in which to conduct its own household economy." Carver goes on to ask, "Who will deny that these moralities function most easily within the privacy of an individual home and are less indigenous to the apartment house?" (1939, p. 271).

In 1963, Carver wrote an ode to the family in seven scenes called "A House is a Place for Flying Apart":

> A house is a machine for living in
> with its pipes for bringing in fresh water
> and for removing waste
> its climate controls and mechanical equipment
> for making meals and entertainment.
> It is also a more subtle kind of instrument
> containing the forces and moods,
> the straining activities and the private tranquilities within a family
> to grow together
> and to grow apart. (1975a, p. 133)

However, in "Planning Canadian Towns," Carver also wrote that "the single-family home in private ownership is not in reality such an essential part of the urban scene as we are sometimes led to believe" (1939, p. 117). Carver argued that the single-family house "isolated on its own private property" is a relic of rural society not suitable to an industrial urban society:

> In a country which is in the process of graduating from a primarily agricultural stage into a fully urbanised industrial state, it is natural to find that the single detached house predominates over all other types of dwelling, since this is the kind of home inhabited by the isolated farm household. British Canadians have brought some of their rural habits of living into town with them and to most of them there is still a special sanctity attached to the single house standing on its own piece of property. (1939, p. 130)

In a "more mature stage of urbanisation," Canada should turn toward "a civic way of living." In urban society, the house should not be taken as a "complete and independent social unity," but part of a larger group of dwellings (Carver, 1939, p. 130). Here, after the metaphor of surgery, we see the urbanist as composer of a landscape of houses. The notes towards "Planning Canadian Towns" would come to form the crux of Carver's later book *Houses for Canadians* (1948), in which he argued for the rationalization of house building, as the system of homeownership in the pre-war period perpetuated the "small-scale, house-by-house method of building cities" (p. 140).

Carver also advocated for rental dwellings as the "basis for planning the city" because the focus is not on the "single house standing on its own piece of property, but on a group of dwellings or neighbourhoods, introducing the "opportunity to plan such groups and neighbourhoods in a scientific manner so as to give better service and pleasanter surroundings for less cost" (1948, p. 140). Carver was particularly interested in Ebenezer Howard's garden city model because the land was acquired as a whole to build the community, and the ensuing rate-rents would be used for and by the community; it formed the basis for his "town centres" proposal in *Cities in the Suburbs*. In 1965, Carver already saw that one of the "central problems of town planning" in the future was going to be addressing the enormous cost of people's desire to live in their own single-family homes.

The contrast between single-family homes and the lesser alternative of high-rise living in Carver's thinking illustrates both his conservatism and his paternalistic interest in improving the lives of the poor and suburban living environments more generally. Even in Czechoslovakia,

Figure 5.2. A typical post-war single-family house designed by CMHC. Source: Bettie Burnett

the garden city movement of the 1920s set an important precedent for the much-admired houses and villas built in Ořechovka and the small family houses built for the workers and their families in the Baťa town of Zlín. Of course, life in those houses came under increasing criticism from leftist thinkers like Karel Teige and Jiří Voženílek, who advocated for communal dwellings over family houses – Carver would be the last person to advocate the break-up of the nuclear family. Carver sought to accommodate both forms of dwelling.

Community Planning: Early Post-war Suburbs

The neighbourhood strategy, as Dolores Hayden (1984) describes it, is based on the attempt to balance privacy and community while also providing for communal facilities. Here, the isolated, rural type of home gives way to communal social institutions more appropriate to urban life, such as "the shared greens, courtyards, and arcades" or the "village and the cloister" that Hayden sees as the model for a neighbourhood strategy that was, in part, feminist in form, with cooperative cooking and eating facilities. She argues that American feminists of the late nineteenth century influenced architects and planners in the United States and Europe, including Ebenezer Howard and Raymond Unwin, whose Garden Cities included the cooperative quadrangles (discussed in chapter 3), where groups of households shared common facilities. In its importation into the United States, the cooperative quadrangles and their communal labour were largely ignored and forgotten, although the idea of a communal green space and courtyards were preserved in the superblocks first implemented at Sunnyside Gardens.

Carver's interest in building neighbourhoods rather than the piecemeal construction of individual houses stems from a number of influences from both the United States and Britain. Carver knew Raymond Unwin as well as Clarence Stein. One of the major influences on Carver, however, and "the guiding principle for Toronto's suburban planners" (White, 2016, p. 93) was Clarence Perry's neighbourhood unit.[5] In his contribution to the LSR's *Social Planning for Canada* ([1935] 1975), Carver writes that "socialist housing plans could be built up on the principle of what may be called the 'Neighbourhood Unit'" (p. 462). The description goes on to summarize the key aspects of Perry's concept, although no reference to him is made. "Planning Canadian Towns" similarly does not acknowledge Perry and describes in far greater detail the characteristics of Perry's neighbourhood unit. Carver was clearly drawing on his work when he describes a "typical Neighbourhood unit" as 160 acres of land (the same as Perry),[6] supporting a population of approximately 5,000 people (like Perry), and offering 10% of the land for recreation (as did Perry) (1939, pp. 213–14). Here the communal space is first and foremost a space of recreation and relaxation. Carver had high hopes when it came to these spaces:

> Every neighbourhood requires what the Russians have so picturesquely described as the "Park of Culture and Rest." The townsman whose life is enclosed within four walls and who is confined by the short perspectives of daily routine, is here released in to the broad reaches of open landscape where Elms, Maples, Beeches and Oaks cast long evening shadows over the green slopes or snow banks. (1939, p. 220)

Although the neighbourhood unit was first and foremost on Carver's mind, the Russian example suggests that already in the 1930s Carver was trying to formulate the neighbourhood unit as not simply a grouping of houses, but part of a larger project to reimagine community beyond a group of single-family homeowners, and with his socialist work at the forefront during this period, it is not surprising that he looked for alternatives in the Soviet world.

Perry's work proved influential in the formation of community planning in Canada. The Community Planning Association of Canada/ Association Canadienne d'Urbanisme (CPA) was founded in 1946, and was funded through CMHC, through part 5 of the 1944 National Housing Act, entitled "Community Planning and Research," which aimed to "encourage public interest in community planning" (Carver, 1975a, p. 88). In these early post-war years, the CPA was governed by a cadre of experts – architects, engineers, and bureaucrats, all of whom were male – designing neighbourhoods.

From 1947 to 1951, the CPA's publication was entitled *Layout for Living* (the name then changed to the more neutral *Community Planning Review*). Urban historian Stephen V. Ward notes how the choice of terminology – "community planning" over the British "town planning" and the American "city planning" – was part of an effort to distinguish Canada's approach to planning from these other traditions, which were still exerting a profound influence on Toronto's post-war planning. (Ward does not mention that the French term chosen was *urbanisme,* placing it clearly in the radar of Le Corbusier's modernist urbanisms.) However, Carver still placed community planning firmly within American and British traditions. In a 1956 memo to CMHC branch managers in Ontario and Quebec entitled "Community Planning," the author (likely Carver) writes that community planning is the Canadian equivalent of British town planning. The term was probably Carver's invention as it appears in "Planning Canadian Towns," and already in this manuscript Carver details the intellectual and planning origins of community planning in a chapter entitled "Evolution of Community Planning." Here Carver lists British (John Nash, Robert Owen, Raymond Unwin, and Patrick Abercombie), American (Daniel Burnham, Robert Moses, the TVA, and Henry Wright), and French (Haussmann and Le Corbusier) influences.

One founding idea of community planning was that some level of industrialization was necessary for the building of houses on a larger scale. As in post-war Czechoslovakia, architects and planners in Canada had to address a building shortage: in 1944, the *Housing and Community Planning* report estimated that a minimum of 730,000 housing units would need to be constructed between 1946 and 1956 (Sewell, 1977, p. 20). Toronto's share of that would be 50,000 units (Carver, 1947, p. 2). In *Houses for Canadians* (1948), Carver would lay out a plan for these 50,000 dwellings by suggesting the creation of 25 new neighbourhoods, each with a population of 7,500 people, a number considerably higher than Perry's 5,000 (pp. 39–40). Carver also drew on another of Perry's influential texts, *Housing for the Machine Age*, shifting his focus from individual houses to the neighbourhood as a whole. Carver describes community planning in the following terms:

> It may be compared with the designing of the process by which the component parts of automobiles are delivered to the assembly line in a rational sequence so that the finished products can be brought to completion as economically and rapidly as possible. (1948, p. 39)

Carver's description reiterates Perry's point that the automobile is "advanced" because it uses the latest production methods and is built by a handful of powerful corporations, while housing is "backward" because there are no large-scale building corporations to rationalize and standardize the production of both houses and the neighbourhood unit.

But in an article for *Layout for Living*, Carver writes that although the industrialization of house construction was necessary, "the production of houses can never quite be like the factory production of cars" (1947, p. 2). Carver shifted his attention from the house as a product to be manufactured on an assembly line to the neighbourhood as a whole: "the individual house is itself only a part of a larger whole. The 'end-product' is not the individual housing unit, but the total community – complete with all the services and utilities which enable urban householders to live as we are used to living" (1947, p. 2). In moving from the house as end-product to the community as end-product, Carver situates community planning firmly within a modernist approach to the city that called for producing space as a whole rather than just individual buildings.

Community planning refers to both the quantitative – the sheer number of houses, schools, shops, etc. – and the qualitative: "Community Planning would be a dull business indeed if it could be justified only on the inhuman grounds of production efficiency" (Carver, 1947, p. 6). Carver argued that the key element of "civic design" means different forms of dwellings arranged such that a community "comes to possess beauty and dignity." As such, Carver writes, "efficiency and beauty" can have for one another a "natural affinity" (1947, p. 6). As a planner, Carver favoured the rental dwelling because it offered a better basis for "planning and building a city" (1939, n.p.). The apartment building, moreover, held no meaning in and of itself, unlike the family house, and as such needed to be incorporated into a larger plan or arrangement. Community planning and its emphasis on "neighbourhoods for

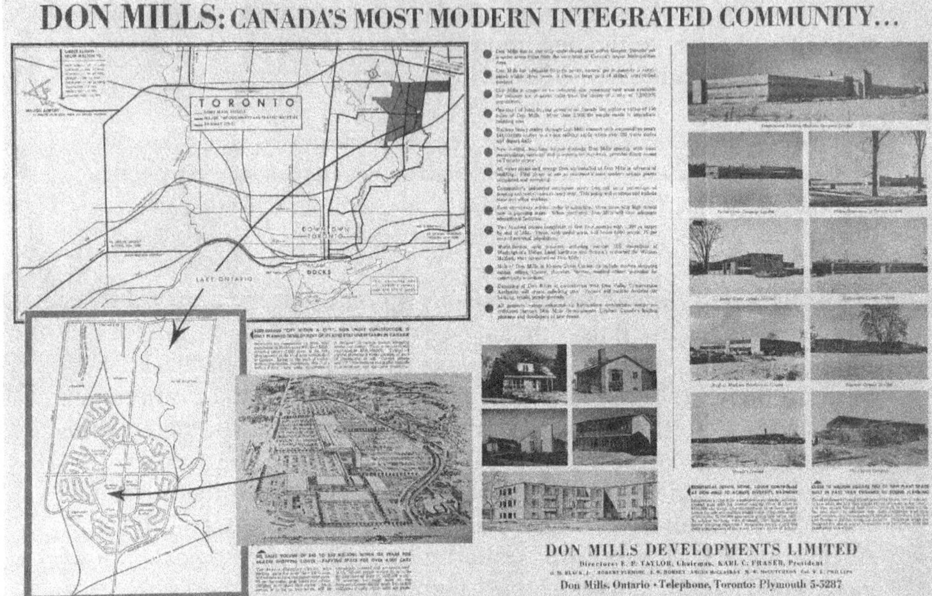

Figure 5.3. Detail from advertisement by the Don Mills Developments Limited promoting "Canada's most modern integrated community." Source: Don Mills Developments Limited (courtesy of Toronto Public Library)

living in" mirrors the early socialist-realist developments, as some of the first attempts to consciously produce space as a whole on the urban periphery drew on principles of modernist urbanism, especially the neighbourhood unit.

One of the key projects of community planning in the early post-war period was the Toronto suburb of Don Mills, lauded as the first "fully planned corporate suburb" – that is, one of the first developer-led, fully planned suburb (Harris, 2004, p. 138).[7] It was Canada's version of Levittown. It developed the neighbourhood unit by grouping four neighbourhoods (for a total population of approximately 29,000) around a central town centre designed by John C. Parkin (see figure 5.4), the architect who would develop the first visions for North York's downtown. An inner ring road would separate the neighbourhoods of single-family houses from the apartments, offices, and shopping centre. It combined rental housing, which also included state-subsidised rentals of both apartment and semi-detached units, and owner-occupied single-family houses, usually bungalows: more than half of the 8,121 dwellings built

Figure 5.4. The Don Mills town centre, designed by John C. Parkin. Source: Courtesy of the Canadian Architectural Archives (Panda Associates Fonds PAN 55943–19)

in Don Mills were apartments (Sewell, 1993, p. 90). However, the development's success and notoriety pushed house prices up, making the neighbourhood largely middle-class. A number of industrial sites planned and developed both north and south of the development were, like in Howard's garden city, to serve as workplaces for the working-class residents, but in the post-war time of increased mobility and combined with the rising house prices in Don Mills, local residents held only 5% of the jobs (Sewell, 1993, p. 93).

It could be said that planning and community are antithetical, but only if community is understood in its now more common understanding as a specific group of people with a shared identity. Community has other meanings as well: common ownership, a community of equals, a neighbourhood. These suggest that community can be planned, but they also suggest multiple narratives for suburban belonging: there was more than one way to create a suburban community and Carver attempted to forge that community out of both his devotion to family life and his socialist leanings.

Cities in the Suburbs, 1955–67

Carver, however, wanted to move beyond the neighbourhood unit as both an organizing principle for urban planning – Don Mills, he thought, was a good start – that would, in part, help bring centrality and density to the suburbs, and as a model for a social democratic vision that would in part offer an alternative to the socialist vision of the sídliště.

Carver therefore committed himself to the art of urban planning in search of a larger specifically urban vision through which he could reconcile his love of houses and family life with his communal ideas around apartment and house dwellers sharing the same community. Throughout his life, Carver likened the work of architecture and urban planning to producing a work of art, whether that was his goal to build beautiful houses or to plan entire neighbourhoods. In the foreword to "Planning Canadian Towns," he writes:

> There have found their way into this book certain ideas, considerations and expressions which may seem inappropriate to the matter-of-fact business of operating a Canadian municipality. For these we make no apology. The building of a city is Man's supreme work of art; and works of art cannot be explained on accountants' balance sheets or calculated on the engineer's slide rule. (Carver, 1939, n.p.)

He made this explicit in his approach to urban space as a work of art. In a 1966 letter to W.J. Withrow, the director of the Art Gallery of Ontario, on the occasion of the City Now exhibition, Carver gave architects a daunting task in his definition of a work of art: "a piece of mega-architecture ... composing the large-scale solids and spaces, weaving into them the dynamic elements of communication and conceiving the whole within a contemporary framework of social philosophy"(Carver, 1966).[8] He was not interested in the photographer capturing images of everyday city life, but in the conscious production of the city as a work of art; essentially, his point was that any exhibition on art in the city necessarily should include urban planners and architects. In his letter, Carver draws links between the cities of old and the new towns: just like Rome and Paris, Cumbernauld and Reston, Virginia were not simply "picturesque accidents," but the "most difficult work of artistic endeavour." Reston and Cumbernauld are significant new towns of the post-war period, but also significant for their well-defined centres. The new towns in the United States, the United Kingdom, and Sweden were important to Carver's work in the 1960s and his research and writing of *Cities in the Suburbs*; they elevated his work beyond the small-scale developments of community planning to developments with strong centres.

One attempt to implement the apartment model of the neighbourhood unit was the Flemingdon Park suburb, not far from Don Mills (shown in figure 5.5). Also a private development like Don Mills, it offered a range of apartments on an exclusively rental basis (White, 2016, p. 130). Macklin Hancock, the landscape architect for Don Mills, produced the master plan, and in order to convince the Borough of North York to approve it, he sent the reeve of the borough and a group of city councillors on a trip to Stockholm and London, where they toured Vällingby and Roehampton, another noted modernist apartment block.

Although Flemingdon Park was built by an American developer, it embodied the post-war ideals of Carver and CMHC. In 1955, CMHC established the "Development Division" and the "Advisory Group" to deal with the "creative" elements of the corporation: research and education on materials and techniques of construction, housing design, and community planning (Carver, 1975a, p. 135). For Carver, the change at CMHC signalled a move away from simple "suburban mortgage lending" to addressing the city as a whole (1975a, p. 136). Carver became the chairman of the Advisory Group, and he remained in that position until his retirement in 1967; he describes these years as "the most constructive part of my working life" (1975a, p. 149). Community planning would be one of the pillars of the group, along with architectural design, building construction, economics and finance, and the vaguely worded "social satisfactions."

Carver wrote *Cities in the Suburbs* while on a one-year sabbatical from CHMC during this period. Although the book is one of any number of early critiques of the architectural uniformity of mass suburbia (cf. Mumford, 1961), the book also resonates with the themes of CIAM 8. The conference's co-organizer, Jaqueline Tyrwhitt, was Carver's colleague at architectural school in London, and the two subsequently kept in touch. In his autobiography *The Compassionate Landscape* (1975a), Carver writes that Tyrwhitt's teachings helped give rise to "much of the idealism for the cities of the postwar world" (p. 24). CIAM 8 had also included two Canadian CIAM groups: those of Ottawa, represented by Alan Armstrong, the community planning expert in Carver's Advisory Group and the first director of the CPA, and Vancouver, represented by Peter Oberlander, who worked with Carver at CMHC. Carver and Oberlander wrote a report to the Royal Commission on the Arts, Letters and Sciences (1949–1951), chaired by Vincent Massey, drawing together the arts, city planning, and advocacy for city planning programs in Canadian universities, of which there was none in the late 1940s – Oberlander

Figure 5.5. Flemingdon Park. Source: Steven Logan

would go on to start the School of Community Regional Planning at the University of British Columbia with funding from CMHC (Oberlander, 1998). The Massey Report, as it is commonly known, claimed that more any other country, Canada's architecture was affected by "mechanical mass production" as so little of its regional architectural heritage remained in comparison with Britain and the United States, which still retained some of the features of past architecture. In comparison to these countries, examples of Canada's architectural heritage are "fewer and the tradition far weaker."[9] Mass production may have completely dominated post-war Czech urbanism, but Czech cities still had the architectural heritage of the previous centuries. The larger issues, for the authors of the Massey Report, were first, a lack of awareness of the importance of good architecture, particularly with respect to public buildings, and the need to foster this through open competitions, and second, the need for the Canadian government to "recognize the importance of community planning and aid it" along the lines of the "Greenbelt Towns" in the United States.

Just as CIAM's 1929 meeting was central to Teige's *The Minimum Dwelling* ([1932] 2002), the core of city and suburban life developed at the 1951 CIAM meeting is central to Carver's visions in *Cities in the*

Suburbs (1962a). Whereas the idea of city cores and the later megastructures were to "counter the appeal of suburbia by offering attractive new environments within the existing city (Gold, 2006, p. 113), Carver attempted to bring the idea of the core into the suburbs. Although he makes no direct references to CIAM's 1951 meeting, Carver seemingly refers to the group's work on a number of occasions, writing, for example, that "Le Corbusier and the advance-guard of European architects were the first to rediscover the 'core'" (1962b, p. 110). In his 1965 book project, he describes "three scenes" in the city: "the Core," "the old city," and "the suburbs." The aim of the core is to create some "civic design" in the commercial centres of cities (Carver, 1965). In *Cities in the Suburbs*, he writes that the suburban explosion has been matched by an increasingly powerful and concentrated downtown, and although it is "the core" of the city, it is no longer "the heart of the city" (Carver, 1962a, p. 7). The core of the city has turned into the "control centre for the new public and private bureaucracies," and this "tremendous upheaval" has caught "the art of town planning unprepared" (Carver, 1962b, p. 59).

The tension between urbanity and mobility at the centre of debates on the core and the megastructure (see Gold, 2006), was also central to Carver's visions of cities in the suburbs. Carver called the building of office towers, shopping centres, and highways the "art of anonymity," the function of which is to be neutral, anonymous, and conforming (1962a, p. 117). He also associated that mobility with the mass-produced suburbs, as well as skyscrapers and apartment buildings, which evinced a lack of "symbolic representation." In an article written for the United Church of Canada in 1967, Carver describes the urban landscape as "expressionless," "anonymous," and "unsymbolic," the "genuine natural environment of a mobile industrialised society." The "ultimate efficiency" of automobiles and expressways and the "glistening efficiency" of the downtown skyscraper is an art of a "cold and compromising kind" (Carver, 1967, n.p.).

Carver saw new forms of communication clashing with the planning ideas he espoused, and this furthered his resolve that what was needed was a core to give form to the sprawling suburbs, a magnet that would attract people. The mass production of the single-family house that began in the late 1940s – and the government-supported mortgages that allowed families to buy houses – along with the changes in urban transportation, from the streetcar to the automobile, had scattered throughout the urban periphery the "bits and pieces and functions of a city" (Carver, 1962a, p. 7). Here, he sided with Lewis Mumford's assessment of urban explosion in the *City in History* (1961), in which he connects the "exploding universe" of the car to a city that had "burst open and scattered its complex organs and organizations over the landscape" (p. 34).

Carver also acknowledges that one of the failings of post-war, mass-produced suburbs like Levittown was the lack of a diverse stock of housing, both of the owner- and tenant-occupied variety, to accommodate a diverse population: young and old, families and single people, rich and poor, etc. If the "standardized material" of the suburbs is to be made into a "work of art," argued Carver, then it needs "variety, surprise and contrast" instead of the row upon row of sprawling family houses (1962a, p. 16). Carver describes the unwritten laws of post-war suburbia thus: "No kind of building but a family house shall enter here. No apartment houses for young people or flats for old people. No corner store. No housing for those who are outside the privileged circle of home-owners" (1962a, p. 16).

Building on Victor Gruen's visions for "shopping towns" (although he never actually references Gruen), in *Cities in the Suburbs*, Carver argues for a network of "suburban Town Centres" at key junctures in the suburbs; these would help to contain the explosion of post-war urbanization and offer a focal point and gathering space for suburban neighbourhoods. In concrete terms, Carver envisions suburban town centres for every four or five neighbourhoods of 5,000 residents each, with a centre serving roughly 20,000 to 25,000 residents, which Tyrwhitt, in her scale of differently sized cores, calls the "TOWN or URBAN SECTOR," the "smallest unit that, in the western world, can be socially and economically self-sufficient" (1952, p. 104). This type of core, within the "urban constellation" of cores of different sizes, usually has a "civic character" (Tyrwhitt, 1952, p. 104). The city centre has four parts: a marketplace, a place for performances and education, the seat of government, and finally the church, which according to Tyrwhitt deserves a "special place in the arrangement of the city" (1962a, p. 96). The government will purchase ahead of time the land that would become the city centre, as was the case with Vällingby in Stockholm. In a 1968 speech reflecting on *Cities in the Suburbs*, Carver outlined this as a key principle of community growth: "increases in land value which occur as the result of the community's growth should accrue to the benefit of the whole community, rather than end up as the private profit of land speculation" (1968, p. 12).

Clustered around the city centre would be apartment buildings for the young and the old, and for people who could not afford homes. He describes the lives of apartment dwellers who take transit as "dependent" and "incomplete," and so they need to be close to the city centre (Carver, 1962a, 18). The single-family household, "a self-contained, independent operational unit," could be located much further from the town centre (Carver, 1962a, p. 18). Although Carver advocated a diverse stock of housing, he still saw the house as the ideal place for families to live, in this way reflecting Clarence Perry's approach to the neighbourhood unit.

The tension between mobility and dwelling in post-war suburban building is related to technologies – cars, televisions, and radios – that serve an "at once mobile and home-centred way of living: a form of *mobile privatisation*" (Williams, 1974, p. 26). Mobile privatization, writes cultural theorist Raymond Williams, is marked by two "paradoxical yet deeply connected tendencies of modern urban industrial living: on the one hand mobility, on the other hand the more apparently self-sufficient family home" (1974, p. 26). Privatization in the form of the isolated family home brought "an imperative need for new kinds of contact." The car allowed for jaunts to the countryside, while through the television people could access the wider world of news and entertainment from the privacy of their home (Williams, 1974, p. 27).

In his notes for an aborted book project in 1965, Carver claimed that the mobility of the population was "one of the worst features of city life" as it meant that people were too "restless" and "unattached" to form coherent, stable, and permanent communities (1965, p. 9). The new transportation and communication technologies, along with the "vehement dedication to home ownership" and the single-family house, were the "anti- nucleation" influences of a mobile-centred way of life (Carver, 1962a, p. 67). Like the paradoxical tendencies of mobile privatization, Carver's own opinion on homeownership is conflicted, if not contradictory: he acknowledges the house gives a family a sense of autonomy and independence, but it also becomes a "self-contained island" and an "anonymous part of the great telecommunication system" (1962a, p. 67). Because houses allowed families to isolate themselves they also discouraged any concrete forms of city organization. Carver writes that "the city is an abstract continuum … without recognizable shape or focus … in which individuals float in a kind of unattached space" (1962a, p. 68). Writing in the pre-Internet age, Carver suggests that "ubiquitous mobility and telecommunication" was turning the city into "a universe within which everyone is in the immediate presence of everyone else," but which was "depreciat[ing] the value of local community" (1962a, p. 67). Within the context of post-war suburbia, where cars, houses, and appliances are "ephemeral, disposable, mortgageable, replaceable, exchangeable" (1978a, p. 5), Carver pleads with the reader of *Cities in the Suburbs*, "can we leave nothing permanent behind?" (1962a, p. 75). In his preparatory notes describing the purpose behind the title, he writes:

> The stuff of which cities are made is scattered in pieces and fragments through this expanse. Can we somehow arrange their pieces so that the

> new "cities in the suburbs" will be triumphant in their comparison with the dignity and excellence of the finest cities of other ages? (Carver, 1962c, n.p.)

In these notes, Carver writes that, along with the family, the single-family house may form a "sacrosanct, closely-knit and internally responsible unit," but there is no corresponding image for the larger neighbourhood of houses; "the mass result of a large number of these houses … expresses nothing in particular" – at least to Carver – aside from the triumph of industrial and construction technology (1962c, n.p.).

In his suburban town centres, Carver seeks to construct an art of living and dwelling outside of the dictates of technological and corporate efficiency, to reproduce the neighbourhood unit in a more collective and communal way, and not simply as a collection of single-family houses. Carver believes that "true artistic expression" can only be found in places concerned with the "meaning of life itself," which did not include alienating forms of bureaucratic labour (1962a, p. 117). Carver wants to emulate the timelessness and monumentality of the churches, cathedrals, and squares of Europe. His suburban town centre is freed and strictly separated from the efficiency and the "practical engineering approach" that has "blotted out any opportunities for excellence in the modern city" (Carver, 1962a, p. 117). He wants to understand and in effect offer a vision for the suburbs that was not defined by its arteries and expressways, nor by homeownership and modern forms of work. This was also how he distinguished the Canadian, democratic approach from the Soviet approach to building linear cities, like Magnitogorsk, in which the worker lives next to the factory, the two separated by a green belt: "this model plan expresses the subservience of the worker to the machine … We prefer to recognize a man politically, not in his capacity as a worker, but as a private citizen" (Carver, 1941, p. 35). In his reference to Soviet life, we can perhaps glean a different interpretation of Carver's attempt to imagine the communal life of the neighbourhood while still preserving the sanctity of private property. He knew about Vällingby and other Swedish examples, and it was with *Cities in the Suburbs* that he sought to replicate this third way for Canadian cities. This would become a key aspect of Carver's attempt to seek out alternatives to the suburb. He continued to think about this point after he had retired. Reflecting on a life spent planning suburbs, Carver (1979) questions the suburban desire to separate working life from living life. Should citizens first be recognized as workers within an economic system, and hence as fundamentally equal, or should the focus be on the pursuit of the individual consumer and aspiring homeowner? Carver sought a third way, one that could be rooted in ideas of community and equality, and yet still emphasize individualism and freedom of movement.

Carver's view of the suburbs, and especially his cities in the suburbs, was essentially an extension of the ideologies of the single-family home into the wider space of the suburban centres. Henri Lefebvre examined these ideologies in his work on the French *pavillon*. The utopian nature of this single-family housing form, according to Lefebvre, stems from the fact that it is defined and understood outside the world of labour and work, which is "put into suspension" (quoted in Stanek, 2011, p. 127). This of course contradicts the way labour and dwelling are so intertwined in the sídliště, particular with the pre-fabricated dwellings: in places like South City, no attempt was made to hide their pre-fabricated construction. Labour and dwelling were intimately intertwined in the sídliště. Stanek writes that the approach to nature in the pavillon was similarly constructed along ideological lines: whereas socialism and its crane urbanism created wide open green spaces, the focus of the pavillon was the private garden, "a mixture of reality and illusion" (2011, p. 127). The green spaces of the socialist city were connected to the overall socialist approach to the city, going back to Moscow's Green City (Bittner, 1998). However, nature in these developments was not simply a matter of green spaces: the socialist city, particularly in Czechoslovakia, also privledged access to the mountains and other places of leisure outside the city, but this was meant expressly as a way of rewarding workers for their labour in the factory, and in many cases, workers would vacation in spas and hotels run by the factory. In the capitalist suburb and in capitalist urbanization more generally, particularly as envisioned by Carver, the spaces of nature outside the city are articulated through the erasure of work and labour, in a return to primordial nature – a return only made possible by the car.

Only Visions without Cars

> The universal possession of a family car and a separate family home is the ultimate privilege conferred by membership in our kind of industrial democracy.
>
> – Humphrey Carver, 1962a

The automobile held a central place in Carver's suburban visions, whether through the neighbourhood unit or the later cities in the suburbs. At the centre of Carver's value system was a particular relationship to nature that was inseparable from the supposed freedoms of the automobile and a necessary component to Carver's understanding of the urban region as a whole. In *Cities in the Suburbs*, Carver sought to update both Howard's garden city and Stein and Wright's Radburn plan. In his preparatory notes to *Cities in the Suburbs* he claimed that the "failure" of the Radburn plan was turning the house around so that its back faced the street, which for Stein and Wright was one of the key aspects of the superblock. By contrast, Carver believed, not that street life should be accorded more importance, but that the "route of approach taken by a car cannot, in fact, be regarded as the back ... Life and liveliness revolves around the family car as a possession almost as important as the house itself" (1962c, n.p.). The ownership of a family car conferred the mobility, autonomy, and individualism that, according to Carver, the non-car-owning apartment dweller lacked.

Carver echoes architect Robert Stern's (1981) evaluation of Stein's work, which Stern felt treated the car as a "problem" rather than a "virtue" by "regard[ing] the relationship of the car to the house as one to be hidden and subverted" (p. 11). Radburn downplayed the car even though it was billed and often remembered as a "town for the motor age" (Stern, 1981, p. 41).

The reworked relationship between dwelling and recreation is central to Carver's vision of *Cities in the Suburbs*, as it was for visions of socialist urbanism in Etarea and South City. There, the architects sought to combat the weekly exodus from the city by intertwining dwelling spaces with green spaces. Carver asserts that the automobile had changed the relationship between city and garden too profoundly for Ebenezer Howard's designs to continue to be relevant. Howard's garden city was unique in that it allowed residents to reach green spaces by foot or on bike. For Radburn, which was not surrounded by a green belt, Stein believed that the "living green close to homes in the midst of the superblocks" was "more essential than greenbelts" ([1950] 1966, pp. 67–8). Carver believed that cars offer access to an "infinity of open country outside the city" and so self-sustaining satellite cities in the age of automobility are unnecessary because "free-ranging travel on regional parkway systems" can easily connect the suburban home dweller with the open countryside (1962a, p. 60). For Carver, the car was an object of liberation from the built landscape, and hence its possession was deemed so important.

In the context of the vast spaces of Ontario's hinterland, Carver offers a similar vision of the relationship between urban space and the countryside, but he premises it on its separation from dwelling spaces rather than its interpenetration. For Carver, the greatest benefit of automobility is its ability to connect people to nature, an idea of nature that Alex Wilson develops and critiques at length in *Culture of Nature* (1991). For Wilson, nature understood as a space of recreation is inseparable from the system of automobility that allows people to access these places. Wilson nicely captures the relationship between nature and technology found in Carver's thinking: "the love of nature flourishes best in cultures with highly developed technologies, for nature is the one place we can both indulge our dreams of mastery over the earth and seek some kind of contact with the origins of life" (1991, p. 25). Nature in this sense is always exalted and never ordinary.

Suburb and nature fundamentally changed with the expansion of automobility. The idea of escaping the city to a recreational landscape is inseparable from the system of automobility that developed in tandem with the idea of nature and with the very ability to get to those places. Carver writes that the Muskoka cottage country outside of Toronto can be as much a part of the city as its downtown financial district (1962a, p. 54). At the same time that the automobile and its highways would separate dwelling from nature, it would paradoxically act to unite city, suburb, and landscape, thereby solving the problems of "metropolitan unity" (Carver, 1962a, p. 58). Rather than see each as separate entities, Carver believes they should be thought of as "all-inclusive regions containing both the city and its *outlying possessions* in the woods and on the lakes" (1962a, p. 58; emphasis added). Carver rightly acknowledges that the forests and lakes should be considered inseparable from the city – the latter are often the source of drinking water – but they became part of the region in a particular way. As Wilson describes it, these are sites of leisure "attached to the schedules and personal geographies of an urban society" where urban inhabitants travel on weekends or on summer holidays (Wilson, 1991, p. 26). Hearkening back to the first parkways of the New York City region, which Sigfried Giedion praised in *Space, Time and Architecture*, Carver suggests that parkways and freeways would create a unity between the suburb and its "spacious playgrounds" (1962a, p. 58). The highways would be the bridge between the city and the country, giving urban dwellers a chance to "share with farmers an interest in cultivating the land for crops and fruits and dairy products, as much for their scenic as for the food value" (Carver, 1962a, n.p.). These recreational and agricultural spaces are enveloped and to a large degree created by both urbanization and industrialization.[10]

Like Carver's sense of house lust, nature was something to be "possessed" as an "open space for recreation" or conserved for "future use and enjoyment" through planning parks and highways (1962a, pp. 54, 57). Carver's understanding of nature is inseparable from both the automobile and the single-family home and reflects the modern obsession with controlling nature and the natural world. Nature is something one goes to, that one looks after, cares for, enjoys, and protects, rather than an integral part of everyday cultural life.

Ironically, it is the automobile that for Carver can renew the city and suburban dwellers' relationship with the "open horizon of land and sky" (1962a, p. 48). Carver's experience of nature is solitary but also fragmented, made more atomized by the distances between the dwelling and nature that the car, in theory, overcomes. In many ways, Carver's views echo the depictions in automobile advertisements, in which the urban and suburban dweller's relationship to nature is inseparable from automobility.

Although Carver argued that the city, suburb, countryside, and hinterland beyond should be thought of as part of one urban region, he still believed that "the suburban city should meet the country with a 'clean' edge" (1962a, n.p.), ostensibly so he could continue to have these experiences. The walled city in medieval Europe is often looked at in this way, its walls marking the boundary between city and countryside and mediating the relationship between city and rural dwellers. This was also Ebenezer Howard's view of the relationship between city and countryside: it was accessible to anyone on foot or on bicycle. Carver laments the fact that the sprawl of the regional city has not only eliminated that fine distinction between city and countryside, but has threatened the city's food supply by encroaching on valuable agricultural land in Ontario.

By Carver's own admission, this rather simplistic "clean edge" divide between city and country had already been thrown into confusion with the rapid development of suburbanization and the expansion of automobility. As early as the 1930s, Carver saw the spread of a new kind of suburbanized landscape in which the car was explicitly implicated:

> On account of the mobility and flexibility of modern transport, the suburbanization of the rural hinterland has been enormously accelerated; the city has set up a process of infiltration and "softening" of the country. There is no longer a clear division between Town and Country. Previously it had always been possible to walk out from the gates of the city and find oneself immediately in the country; now the front has become fluid and in between Town and Country there is a wide transitional area which is neither one nor the other. (1939, p. 55)

This paradoxical belonging and separation was magnified by the car, which allowed, and to a certain degree necessitated, thinking beyond the city in a way that encompassed the region as a whole, which included the countryside and recreational regions. Carver premises his vision of nature upon a simultaneous, and contradictory, critique of "ubiquitous mobility" and a celebration of mobility and mastery in the access to the hinterland that the car provides. In a revealing passage from *Cities in the Suburbs*, Carver asks: "What's the use of a car if you can't get to the water, the woods, and the mountains? What's the use of getting there if water, woods, and mountains are not yours to enter?" (1962a, p. 55). Architect James Dunnett offers a similar assessment of the approach to both nature and the car put forth by Le Corbusier in *The Radiant City*. He notes that after visiting New York in 1935, Le Corbusier changed his view of the car: "The new city is compact – the transport problem solves itself. We learn to walk again – the motor car (there are 1.5 million of them in New York every day) is a malady, a cancer. It will be valuable at the weekend or even every day to disport oneself amidst the tender verdure of nature" (quoted in Dunnett, 2000, p. 70). Le Corbusier appreciated the car inasmuch as he could use it to reach the "tender verdure of nature," although over time, with the increasing distance between home and work, it became far more valuable for the commute. The meditative aspects of nature were also central to Le Corbusier's vision of his radiant city. It was not that modernists like Le Corbusier had a love affair with the car, but that his plans made "little provision for the car numerically, even by the standards of the time" (Dunnett, 2000, pp. 68–9). Like in any automobile advertisement depicting the open road, one will never find a traffic jam in an architectural plan of a city. The naivety of modernist architects and planners, like Carver and Le Corbusier, was reflected in their belief that "green cities" were "compatible with the car society" (Wolf, 1996, p. 158). As the separations between home and work widened in post-war modernist urbanism, the car was not only for weekend jaunts, but the daily commute.

The problem for Carver, like many other modernists, is that he was not able at the time to imagine the ways in which automobility would expand, the ways in which the roads would come to be filled with cars on their way out of the city to the countryside. A fitting counter-image for the lyrical praise of the highways to/in nature voiced by Carver, Le Corbusier, and Giedion can be found in the opening minutes of Jean-Luc Godard's 1967 film *Week-end*, in which we see a string of car collisions littering the landscape and the road leading out of Paris.

It signalled the beginning of the end of optimistic thinking about the open road connecting city and countryside (romanticized in countless films as well) and reflects Carver's disappointment with the expansion of automobility.

Did it occur to planners like Carver that the car might in the end deny the very experience of nature it promised? At the age of 92, a year before he died, Carver made the following admission in his final work, the self-published book *Decades*:

> Very early in life I wondered how cities could grow without destroying the surrounding countryside ... I became an advocate of Garden Cities. I wrote books (*Cities in the Suburbs*) and helped to establish organizations (such as the Community Planning Association) hoping to restrain the destructive forces of city growth. But I failed ... The big city, with its network of freeways, has become a monster which swallows up the landscape. That is the major change that has occurred in the 20th century. (1994, p. 97)

As urban inhabitants escape en masse to the surrounding countryside, the countryside is made more difficult to reach, particularly for those who do not own a car, and in Carver's case for those who could no longer navigate a busy highway: "I can't drive at 90 kilometres an hour and read the Exit signs or make quick decisions. I am practically helpless to exist in my own habitat" (1994, p. 97). In light of this realization, and in light of his own dismay at global urbanization, Carver was able to retreat to the world of his single-family home in the well-to-do, leafy Ottawa neighbourhood of Rockliffe Park Village, where he declared "small is beautiful. We like it just the way it is" (1994, p. 98).

Conclusion

In "Building the Suburbs: A Planner's Reflection," Carver (1978b) reflects on three fundamental historical forces that fueled suburban growth: mobility, freedom, and individuality. For Carver, as this chapter has shown, each of these had negative consequences, even if the intentions underlying them may have been noble. Humphrey Carver offered one of the early critiques of the mass-produced suburb, pointing to the lack of an overall vision for a neighbourhood that should include parks, community centres, diverse housing stock for all classes, and good public transportation.

However, even in *Cities in the Suburbs*, in which he makes all manner of architectural and urban critiques of the homogeneity of the post-war suburb, he still very much subscribed to what Dolores Hayden calls the "Victorian template of patriarchal life" and an "outworn family model" of male and female stereotypes (1984, p. 44). In stark contrast to the socialist urbanism of the 1920s, the family was for Carver the core unit of post-war social democracy, as well as a moral guide. Although Carver argued that new appliances in the household "help to liberate the housewife from the monotonous servitude of domestic chores," he did not mean that she could leave the household and pursue other activities; rather, such technological advances would better "allow her to develop family life in more fruitful directions" ([1935] 1975, p. 463). If Teige premises his theories of the minimum dwelling on the rejection of the single-family house and the institution of the family, marriage, and monogamy, Carver makes them the moral and ethical bedrock of his vision.

This kind of approach was in part why Carver all but ignored the feminist roots of the neighbourhood strategy, as he subscribed to the stereotypical gender roles commonly associated with post-war suburbanization in the 1950s. The art of the suburbs was not only about an aesthetic arrangement of space, but also a "programming of space" to suit the "highly normative schemes of human values" (Hayden, 1984, p. 120) represented by Carver's thinking about suburbs, family life, and work. In *Cities in the Suburbs*, Carver paints a picture of "car-borne workers and families with car-borne wives," suggesting that the businesses of the suburban town centre should cater to the "housewife's market basket" (1962a, pp. 98, 82). Carver called one of the major themes of his early post-war work with CMHC "design for living," which upheld family life as the core of the "good life" and the larger neighbourhood and social life (1975a, p. 115). The single-family house situated away from the centre was the most suitable place to raise a family, and access to the car was the key to a family's independence and autonomy.

Carver did begin to question some of his long-held convictions about homeownership and automobiles in the wake of his retirement from CMHC in 1967; this tendency was also encouraged by broader changes in planning, particularly in Toronto, towards which Carver gestured in many of his post-retirement speeches. In the opening lines of a 1979 speech given at the University of Waterloo, Carver explains that "like any old man groping and stumbling in the dark, I have been trying to raise in my mind some picture of the new Habitats or Living

Places that might now begin to emerge in Canadian cities … It might be called a 'post-suburbia' habitat." For Carver, this habitat treated the individual as a social being first, not a car or homeowner: an individual entitled to her own private space while at the same time living in an environment of "social space" in which people of all incomes and classes inhabit – a "social mix" rather than segregated public housing (1979, pp. 1, 14–16).

In a speech to the Canadian Housing Design Council, Carver clearly was thinking of a new approach to the suburb, one based not on the separation of dwelling spaces and the surrounding landscape, but an interpenetration of the two, where anyone could leave their house and ski, walk,or bike along paths, trails, lakes (1975b, p. 7). This was an affirmation of the Scandinavian models like Tapiola, Finland, which had so impressed Czech architects and urbanists on their visits.

Carver called for a "retreat from the universal aspiration" of owning a single-family house with a garden; instead, he looked to more "collective forms of housing" that were part of the landscape and not separate from it: "give up the colonial house and yard as a holy and natural symbol," he argued, and "bring the natural world to our doorsteps in a real and intimate way" (1975b, p. 8). Carver retreated from his love of single-family houses and called for more compact collective forms of dwelling where inhabitants would not rely on cars. Although he was certainly not envisioning the break-up of the family, as Teige and other leftists of the 1920s had advocated, the architects in post-war Czechoslovakia were also considering collective life without the break-up of the family. Carver's visions returned to the garden city principles of collective spaces that North American planners rejected when the model became part of suburban planning dogma, but which Czech architects and urbanists were implementing on the ground in places like Solidarita.

In 1967 – Canada's centennial year – Carver had proposed that a number of "model Town Centres" be built in Canadian suburbs. Combining exhibits with model housing, Carver situated his proposal in the lineage of the 1951 Festival of Britain and various World's Fairs, which always left behind "permanent civic improvements" (1961, p. 3). In a later article, he turned specifically to Yonge Street – "The Main Axis of Toronto" – on which "a series of focal points" can serve the design of "sub-centres": "At St. Clair and Finch are well developed hearts of local communities; these may be prototypes for future centres farther out on the axes of the city's growth" (Carver, 1962b, pp. 59, 62). Yonge and Finch was to be the northernmost starting point for the modernist redevelopment of Willowdale.

Although the model that was followed differed from Carver's vision, Carver's role in developing what by the 1980s would become planning dogma is unacknowledged, as are the multiple ways in which Carver's own ideas were influenced by international planning and the architecture of CIAM. Willowdale was Toronto's first attempt to put Carver's subcentre theory into practice.

Figure 6.1. The clash of forms: the mass-produced single-family house and the condominium tower of the post-suburban landscape. Source: Steven Logan

6 The "Total Image": The Making of Willowdale Modern

At Weissenhof, Germany, in the year 1927 Europe's best architects planned and developed a Demonstration Project of community planning and architectural design ... North York is the appropriate locale for a *new* Demonstration Project for North America ... Inventiveness, daring, advanced ideas and concepts in the Yonge Street corridor, just as at Weissenhof 41 years ago, might become the "norm" for developments throughout Canada, indeed the world.

– James D. Service (Mayor of North York), 1968

The 1927 Weissenhof experimental housing settlement in Stuttgart, Germany, under the direction of architect Mies van der Rohe, brought together 17 architects, including Le Corbusier, Bruno Taut, Mart Stam, and Ernst May, to build 33 houses. Karel Teige called Weissenhof "an event of international significance for the entire modern world" ([1932] 2002, p. 187), while László Moholy-Nagy called it the "most spectacular demonstration in the history of modern architecture" ([1947] 1965, p. 108). Housing colonies as experiments in modernism were common on both sides of the Atlantic: the Prague housing colony Baba and the Brno housing colony were scaled-down versions of Weissenhof, and Invalidovna, constructed in 1967, was an early example of an experimental sídliště using pre-fabricated technology. In the United States, Dolores Hayden points to Clarence Stein's successful "demonstration projects" in Radburn, Sunnyside Gardens, and Baldwin Hills Village in Los Angeles.

On 27 February 1968 at the Inn on the Park hotel, North York mayor James Service gave the speech from which the epigraph that opens this chapter is taken; he was speaking to an assembled crowd of bankers and developers on the occasion of the public unveiling of the Yonge Redevelopment Plan authored by the noted modernist architect John C. Parkin and urban planner Murray Jones (see figure 6.2). The study cost

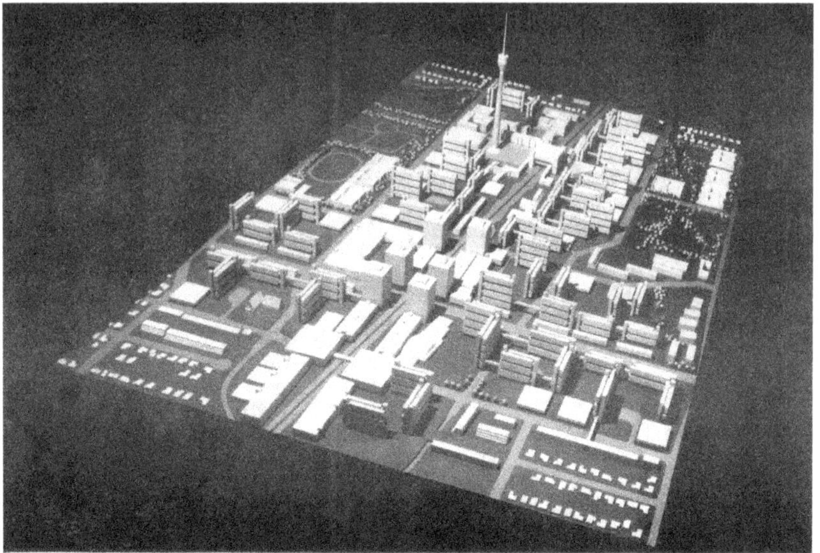

Figure 6.2. Model for the redevelopment of Willowdale. Source: *Yonge Redevelopment Plan* (1968)

the borough $164,000 and was the most expensive and ambitious plan to date in Metropolitan Toronto.[1] Service was hoping the assembled guests would buy into the plan, which over a 20-year period would provide 23,000 new housing units for 60,000 new residents, and jobs for 20,000 people in the new high-rise office buildings built in anticipation of the subway extension into the Toronto suburb of Willowdale.[2] On the day after its release, the headline of North York's newspaper, *The Enterprise*, read, "Sky High Schemes Unveiled," with the article's author boldly claiming: "It's exciting. It's imaginative. It's big. It's probably the only concept of such magnitude in Canada and North America."

Jones and Parkin's plan took Service's comparison seriously, creating a monumental tribute to heroic modernism in the spirit of Le Corbusier. Service's comparison to the Weissenhof housing project clearly placed high expectations on the redevelopment of this suburban strip, which was to become the heart of the newly formed Borough of North York. Service situated Willowdale's redevelopment within the history of modernist urbanism, and rightly so. North York, he argued, was on the avant-garde of planning, bringing together developers' private capital with public funding and support for a range of projects, beginning with the borough's subdivisions and their mass-produced, Levittown-like

Figure 6.3. The sprawling stretch of Yonge Street, late 1960s, where the redevelopment was to take place. Source: *Yonge Redevelopment Plan* (1968)

Figure 6.4. Marrying horizontal and vertical transportation: looking towards the skyscrapers of Willowdale from the south side of the 401, Canada's busiest highway. Source: Steven Logan

houses in the 1940s and '50s, and extending to the "new towns within a city" like Don Mills and Flemingdon Park. Service clearly championed North York as the suburban home of modernist urbanism.[3]

Although the history of Willowdale and the other villages that made up the once-rural township of North York has been well documented ever since suburbanization began (Hart, 1968; Kennedy, 2013), post-war North York had aspirations to be above all a modernist city. In his preface to *North York Modernist Architecture Revisited*, Toronto architect and writer Michael McClelland notes that the post-war period was "the perfect time to explore the most advanced ideas about architecture and city building and to test them out in quickly developing North York" (2010, p. 8).

This chapter focuses on the initial plans, dating from the 1960s, and the subsequent public reactions and plan revisions that took place in the 1970s and '80s. In the controversy over Jones and Parkin's plan we can see the tension between the classical suburb of family houses and the planned density of high-rise office towers and residential buildings. The story this chapter tells of Willowdale's redevelopment reflects Carver's ambivalence about ubiquitous mobility, the encroaching "art of anonymity" engendered by skyscrapers and highways, coupled with his desire for both community space and the individual space of the family home. Willowdale may have been the ideal place for Carver's post-suburbia habitat to take root, and although the end result brought a dense city centre to the suburbs, it may not have always matched the ideals Carver espoused at CMHC and in his personal writings.

The chapter situates Willowdale within Carver's ambition for suburban city centres, while also placing it within the larger international dialogue around bringing density and centrality to the suburbs in the 1960s. Unlike South City, this was not an entirely new development, but the focus on building a suburban downtown and the overall modernist philosophy that imbued the planning in both Willowdale and South City is significant. The architects and planners in both places cited Cumbernauld and Vällingby as inspirations, even though their understandings of the modern suburb and the ways in which their plans would be implemented radically differed.

Vision in Motion

Although the turn to subcentres can be read simply as a history of policy change in Metropolitan Toronto, the philosophy behind it connects to the wider currents of modernist urbanism. Carver's *Cities in the Suburbs* is central in this regard, as is the work of Hans Blumenfeld and Jaqueline Tyrwhitt, both of whom drew on Bauhaus avant-garde artist and theorist Laszlo Moholy-Nagy's ideas, particularly his book *Vision in Motion.*[4] In that book, Moholy-Nagy ([1947] 1965) uses the term "vision in motion" as a catch-all phrase to describe changes in experiences of space-time that united art, the applied arts, architecture, painting, and film and that marked the avant-garde of the 1920s. Vision in motion is a "synonym for simultaneity and space-time" (Moholy-Nagy, [1947] 1965, p. 12) that sees objects, like buildings, not as isolated phenomena, but relationally as part of a coherent whole.

Tyrwhitt's essay for Marshall McLuhan's journal *Explorations*, entitled "The Moving Eye," brings together her interest in developing an idea of the core with Moholy-Nagy's vision in motion. The title of Tyrwhitt's essay comes directly from Moholy-Nagy's book: "A new viewpoint in the visual arts is a natural consequence of this age of speed which has to consider the moving eye" (Moholy-Nagy, [1947] 1965, p. 246). In Tyrwhitt's article, though, she does not discuss the moving eye of the automobile, but rather turns to a place that predates the automobile: Mahal-i-Khas, the core of the sixteenth-century "dream city" of Fatehpur Sikri in India, close to the Taj Mahal in Agra. Tyrwhitt describes the particular way that the visitor to Mahal-i-Khas always feels at the centre of things: "from the moment he steps within this urban core he becomes an intimate part of the scene, which does not impose itself upon him, but discloses itself gradually to him, at his own pace and according to his own pleasure" ([1955] 1960, p. 90). She argues that planners need to "rediscover the importance of vision in motion" (Tyrwhitt, [1955] 1960, p. 94).

Although Moholy-Nagy's vision in motion implied a rejection of the past ([1947] 1965, p. 260), Tyrwhitt turns to a pre-automobile and pre-industrial city for inspiration. Given that CIAM's focus on the core emphasized the rights of the pedestrian, this seems logical. There is also a clear tension between Tyrwhitt's focus on a gathering place for pedestrians as the setting for the "moving eye" and the dominance of the automobile in the early post-war period.

Hans Blumenfeld, however, placed the car at the centre of his visions in motion. In the late 1960s, Blumenfeld also called for a system of subcentres in the metropolitan region, the design of which would respond to and accommodate the growing use of the automobile and other forms of rapid public transportation; his ideas would be explicitly taken up by Parkin and Jones in their vision for Willowdale, and particularly the main axis of Yonge Street. Like the Czech architects and urbanists developing the living environment and the need for "synthesis," Blumenfeld understood the need for the "seeing together of the interaction of all the factors which determine the life of society, its content. It also meant a seeing together of all the elements of the physical environment, as the perception and conception of form" (1967a, p. 305). Blumenfeld called it "together-seeing." Two aspects are important for Blumenfeld in this together-seeing: the new "extra-human scale" emerging with new technologies, such as the automobile, skyscrapers, and freeways, which, in being "extra-human," were part of "'outer' nature like mountains and rivers," and a "system of sub-centres" through which and in which these new elements of the physical environment could be ordered and expressed (1967a, pp. 308, 310). The skyscraper and the elevator – the "product of mechanical means of vertical transportation" – along with the "horizontal extension of the metropolis" associated with the car had "obliterated the street as a defined space of inter-related proportions" (Blumenfeld, 1967a, p. 309). With the street obliterated, the key element in the design of the subcentres of the metropolitan landscape are separate environments for both pedestrians and cars.

In a seeming homage to Moholy-Nagy as well as Kevin Lynch,[5] Blumenfeld writes that for urban design to give form to these subcentres means "unfolding the total image as a sequence of memorable images along the paths of vision in motion" (1967a, p. 309). The total image also meant separating pedestrians from cars, and "shaping pedestrian islands on the human scale and … connecting them with each other."

To design for a "motorized world" means designing the "view to the highway" and the "view from the highway," where "the driver's vision in motion can build up a composite memory image of the metropolis comparable to the composite image which was built up by walking through the streets of older and smaller towns," much like the way Tyrwhitt described Fatehpur Sikri, although Blumenfeld writes that the "system of spaces" in the modern city can no longer be apprehended by walking (1967a, pp. 309, 308). For Blumenfeld, it was the city in motion, a devotion to mobility *as such*. For Blumenfeld the suburban subcentre was irreducibly modern.

The Subcentre Approach

Blumenfeld's philosophy of subcentres had been the focus of both planning and design since Carver's publication of *Cities in the Suburbs* in 1962, and also, within the wider North American planning discourse, Victor Gruen's *Shopping Towns USA* (1960). Pierre Filion (2001) and John Sewell (1993) trace the focus on subcentres in Toronto specifically to Metropolitan Toronto's 1980 Official Plan ("Metroplan"), which aimed to shift downtown office space to the suburbs. This was a key aspect of the "suburban mixed-use centre policy" of which Willowdale, along with the Scarborough Town Centre, would be the two key centres outside of downtown Toronto. In the initial discussions, there were also to be 13 other "intermediate" centres throughout Metropolitan Toronto (see figure 6.5), but this was reduced to four in the official plan (White, 2016, p. 356). According to White, this was part of Metroplan's larger objectives of a "multi-centred" metropolitan area with "mixed-use centres" that would "enrich the suburbs with downtown-style culture," allow for people to live close to their work, and help create an "efficient 'nodal' public transit system" (2016, pp. 356, 352). Sewell (1993) and White (2016) connect Metroplan's focus on subcentres to the policy discussions of the 1960s, and especially the 1962 Metro Toronto and Region Transportation Study (MTARTS), which strongly advocated a regional plan for Toronto addressing many of the concerns Carver was expressing in *Cities in the Suburbs*. It recommended that urbanization should be channelled into "discrete, coherent communities, each with its own downtown," the idea of which, White suggests, likely came from White (2016, p. 228). And for good reason. Carver, along with Hans Blumenfeld, Albert Rose, Len Gertler – all planners – were part of an MTARTS subcommittee charged with "assembling the foundation for a regional plan" (White, 2016, p. 226). In a 1964 letter to CMHC president Herbert Hignett, Carver suggested that the authors of MTARTS

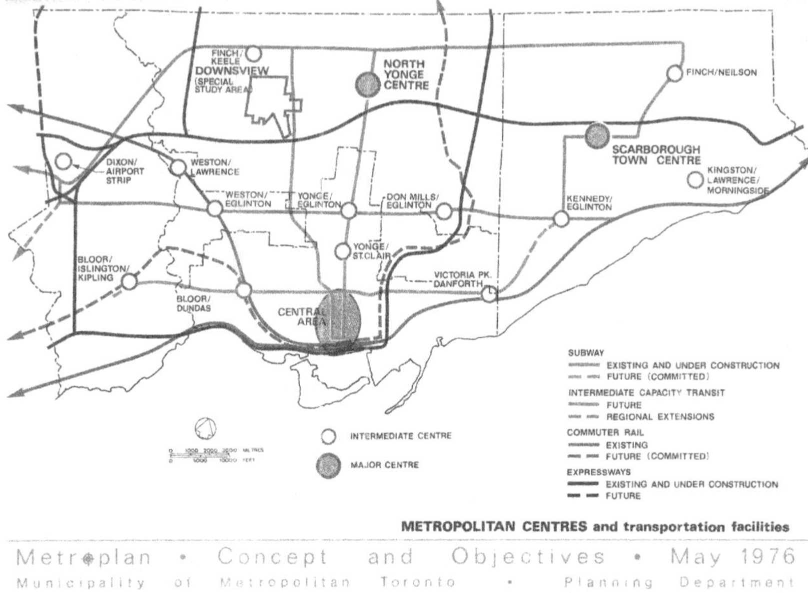

Figure 6.5. Map of Metropolitan Toronto with proposed subcentres. Willowdale is designated the North York Centre. Source: Metropolitan Toronto Planning Department (1976)

asked for his input because they were likely interested in his "suburban town-centres" as "an essential part of the physical scheme of development that would have to be served by transportation" (Carver, 1964). In a note accompanying the publication of a shortened version of the Yonge Redevelopment Plan, Mayor Service writes that the emergence of "regional cores" in Etobicoke, Scarborough, and North York is a "recent phenomenon which … will strengthen Metropolitan Toronto by providing variety, vitality, a visual focus and a social identity to these hitherto 'dormitory' areas" (1968a, n.p.).

The subcentre approach advocated by Carver would take up a central place in *Choices for a Growing Region*, the main publication of the MTARTS planning committee. Among the report's key regional goals were "Environment" and "Accessibility," both of which emphasize the role of subcentres in "facilitat[ing] and maintain[ing] a pattern of local communities," as well as minimizing commuting time between home, work, shopping, and recreation. Residential neighbourhoods specifically should be located so that "they are accessible to the more distant recreational hinterland without fatiguing in-city travel" (MTARTS, 1967, p. 23).

A further study, the 1974 Metropolitan Toronto Transportation Plan Review, proposed the idea of subcentres, or "downtowns in the suburbs," including one in North York (Sewell, 1993, p. 217). By the end of the 1970s, the subcentre approach was "common wisdom," becoming part of Metropolitan Toronto's "Centres policy," its official response to both unchecked urbanization and also an implicit critique of Athens Charter urbanism and its separation of functions (Sewell, 1993, p. 219). The policy outlined that the subcentres should be "multi-functional," "pedestrian oriented," and "intensely developed." Sewell suggests this meant rejecting the "modern idea of separated, segregated uses'" (1993, p. 219), although Sewell's modernism, which he equates with Athens Charter urbanism, does not address the critique that already come from within the modern movement beginning with the 1951 CIAM conference on the core, which called for the kinds of city centres that this policy advocated.

The earliest plans to develop a civic centre in Willowdale actually precede Jones and Parkin's plan and date back to 1963, when North York Township staff prepared a "rough draft" of a plan drawn up by Raymond Skelly, a former chief planner of Edinburgh (Matthew & Davidson, 1983, p. 1), which Service referred to as the "Skelly plan" in his speech (see figure 6.6). Skelly's plan, entitled "Civic Centre for North York," offers in embryonic form much of what would appear in the later plan and reflected the dominant modernist ideas of the time, like the multilevel urbanism proposed in cities and new towns throughout the world.

Skelly's proposal suggested that the only way to keep the centre compact *and* accommodate the automobile is by segregating cars and pedestrians, by keeping "all pedestrian movement on a platform above service roads and car parking" (1963, p. 11). This would be part of a network of pedestrian walkways inspired by the fact that many people in Willowdale live within walking distance of the centre: "every opportunity should be taken to provide continuous pedestrian access from outlying residential areas into the new centre. Connection between pedestrian ways and the pedestrian platform should take the form of a gradual series of intermediate levels linked by ramps and steps to create attractive access to the upper level" (Skelly, 1963, p. 22).

Skelly's plan also began to formulate how to separate existing suburban residential development from the new development and traffic of Yonge Street, which would become one of the redevelopment's most profound changes. Skelly proposed only allowing certain east–west streets to cross Yonge Street, with the remaining streets forming "cul-de-sacs" or "loop roads," which would increase the "intimacy of the residential areas" (Skelly, 1963, p. 13). He proposed a ring road

Figure 6.6. R. Skelly's 1963 plan for Willowdale, which predated Jones and Parkin's plan, made remarkable by the complete erasure of the existing streetscape on Yonge Street. Source: *A study of the existing civic strip on Yonge Street from Sheppard Avenue to Finch Avenue: Incorporating proposals for a new civic centre for North York* (1963)

surrounding the new development and separating it from the existing suburban landscape. The plan is remarkable for its complete rejection of Yonge Street as a place for people to meet and do business, with the demolition of the existing two-storey built fabric, the stretch of Yonge would be dominated by green spaces and residential buildings with a few "genuine highway commercial uses" remaining, like "service stations, tire depots and car-sales showrooms" (Skelly, 1963, p. 12).

Even before Jones and Parkin's plan, Willowdale was already a part of the "modernism thinks big" approach to planning outlined in chapter 2. The multilevel urbanism and clear separation between pedestrians and cars, along with the rejection of the street as social space, was typical of so many modernist projects throughout the world, including those in the Eastern Bloc.

The Plan

Murray Jones and John C. Parkin were two of the most significant actors in Metropolitan Toronto's urban and architectural landscape.[6] Parkin studied under Walter Gropius at Harvard and built many of North York's most noted modernist buildings in Don Mills, including the Bata International Centre (1965), Ortho Pharmaceuticals (1955), Don Mills Shopping Centre (1959), and the firm's own office (1956). Parkin also took part in the realization of Viljo Revell's plans for Toronto City Hall, completed in 1965. North York would attempt to ride this modernist wave with the Willowdale redevelopment. Murray Jones was one of the first leaders of the Metro Planning Department and his deputy was Hans Blumenfeld. Through his firm, Murray V. Jones and Associates, Jones prepared a number of "urban renewal" schemes for cities in Ontario, including Hamilton, in the late 1960s.

Jones and Parkin's Yonge Redevelopment Plan drew on Blumenfeld's work using Willowdale as a test bed for the new scale of cars, subways, and highways. They were also arguing for a new conception of the urban, suggesting that the term "suburbs" was as "obsolete as the term 'city'" (Jones & Parkin, 1968b, p. 2). Jones and Parkin believed that the "metropolitan area," which came about with changes in transportation and communication – streetcars, telephones, subways, elevators, and cars – had led to the "disappearance of the city as it was traditionally known" (1968b, p. 2). For Jones and Parkins, the compact areas of Toronto's earlier settlements were "determined by an economy and level of technology fundamentally different from that which obtains today" (1968b, p. 1). In submitting their plan, Jones and Parkin write that their task was a difficult one because "a theory of urban sub-centre planning has yet to be formulated" and their design would provide the basis for "creating a distinctive element of the total urban structure" (1968b, n.p.).

In the first part of the Yonge Redevelopment Plan, entitled "The Urban Process and Purpose of the Plan," Jones and Parkin explicitly interpret Willowdale's history of dwelling and transportation and the "natural development" of Yonge Street within the language of subcentre planning, showing that villages like Willowdale had by the nineteenth century become "complementary sub-centres" in their own right "providing local services to surrounding farm lands as well as functioning as links in the transportation routes which connected other centres" (1968b, p. 3). Willowdale occupies an important place in the history of suburbanization in Toronto because it grew up on Yonge Street, the main north–south street connecting the city to its hinterlands and dividing the eastern and western parts of the city. Willowdale and the surrounding villages up and down Yonge Street remained largely farming communities well into the 1920s, with some of the rural life and much of its architecture persisting into the 1950s.

Still, subdivisions were already being built in the 1910s, when rural landowners began selling plots to real estate developers keen on taking advantage of the bucolic surroundings and the Radial, an electric streetcar that ran the length of Yonge Street from downtown Toronto to points further north of the city. During these times, shops, inns, and taverns would cluster around the stops of the Radial. By the 1920s, the Yonge Street villages were already considered a "settlement of commuters" who used the streetcar to get to work (Hart, 1968, p. 260). As the remnants of the farming community disappeared, North York's population rapidly increased. In 1948 the population was 38,000, and by 1958 it was 200,000. In 1967, North York Township officially became a Borough of Metropolitan Toronto, and by 1968, when Mayor Service unveiled the redevelopment plan, the population had reached 425,000 (Hart, 1968, p. 302). In 1968 alone, 74 apartment buildings were constructed, and all together 12,000 apartments were added; these numbers are remarkably similar to the numbers for Prague, which was to also build 12,000 apartments annually in the 1971–5 period. It was in this context that Service, elected in 1965, and the North York Council approached John B. Parkin Associates to prepare a plan to make Willowdale the site of a civic centre for North York. In *Pioneering in North York* (1968), Patricia Hart, co-founder of the North York Historical Society, placed Jones and Parkin's plan firmly within Willowdale and North York's development: the model appears next to an 1860 map of North York as the frontispiece to her book.

Although Toronto urban historian Richard White suggests that the American and British new towns had little in common with Toronto's "mixed-use centres within an existing urban area" (2016, p. 357), Parkin and Jones turned to Europe for examples, specifically Stockholm, where new towns and subways were built as part of one operation, which was the way that Mayor Service consistently framed the Yonge redevelopment to both angry residents and potential developers: development would inevitably come with the subway, so better to plan for it. Parkin and Jones referred to the town centres of Vällingby and Cumbernauld. At the same time, they distinguished the Willowdale plan from Vällingby, noting that Willowdale was already "largely settled" and the state owned very little of the land, and as such the economic problems of implementation would be different. This was a plan driven by private investment and the promise of tax revenue from future businesses.

One of the aims behind the plan's concept (shown in figure 6.7) was to "provide for a growing metropolitan sub-centre while also preserving existing areas" (Jones & Parkin, 1968b, p. 5). Although many houses would have to be expropriated in Willowdale's transformation – according to

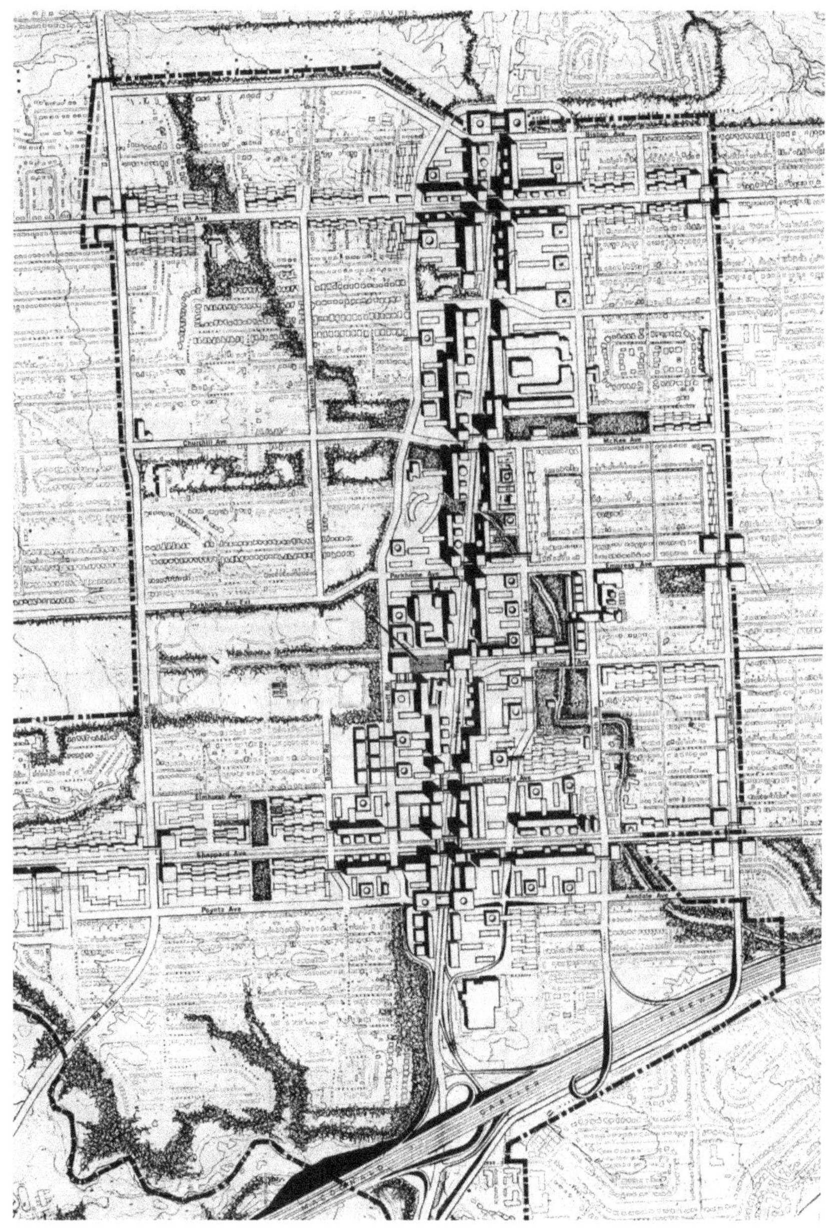

Figure 6.7. Detailed urban plan for the Yonge redevelopment area. Source: Yonge Redevelopment Plan (1968)

The Enterprise, 2,100 "dwelling units" (28 February 1968) – the suburban character of the neighbourhoods beyond Yonge Street from both the interwar and post-war periods were to be preserved, separated from the new downtown by two north–south minor arterial roads running east and west of Yonge Street between Highway 401 to the south and Finch Avenue to the north. These roads – Beecroft Avenue to the west of Yonge Street and Doris Avenue to the East (see figure 6.9) – would channel local traffic away from Yonge Street and their medians and landscaping would offer a buffer between the high-rise centre and the existing single-family-house development. The area in-between would become the "primary corridor of intensive development" (Jones & Parkin, 1968b, p. 5), marking the divide between city and suburb, between high density and low density, between commercial development and high-rise residences and the quiet suburbs and green landscapes, reinforcing modernist urbanism's contradictory goal of uniting urban space through separation, while at the same time making it a sub/urban landscape (Keil, 2018).

In their search for a "uniform philosophy" of "sub-centre planning and realization," Jones and Parkin quote at length from Blumenfeld's article, taking on his challenge to "unfold the total image as a sequence of memorable images along the path of vision in motion" (1968b, p. 2). The passage became a general guide to their design.

Along the paths of Yonge Street, Willowdale would become that "sequence of memorable images," the main features of which would be a 1,000-foot communications tower and a city hall that would straddle the street (see figure 6.8). The sidewalks on Yonge Street would be replaced by a "traffic buffer for landscaping" (Jones & Parkin, 1968b, p. 102) and a place for "automobile-oriented signs and show cases" (North York Planning Board, 1969, p. 45). This set-up could better relate advertising to the new scale of the car. This interest in developing advertising "as an art form at the city scale" (Jones & Parkin, 1968b, p. 112), reflects Moholy-Nagy's interest in turning static advertising into a "kinetic process" aimed at the "rapidly changing position of the spectator at the wheel" ([1947] 1965, p. 246). The cars would pass under the new North York City Hall while also giving pedestrians an opportunity to cross Yonge Street and look down on the traffic spectacle below. In its design it evoked the Cumbernauld city centre that Jones and Parkin cite as an influence, and which also straddles the roadway. These were the new monuments to circulation, the modernist successors to the Arc de Triomphe in Paris.

Although the sidewalks would be completely removed from Yonge Street, an elaborate pedestrian network built at grade, and below and

Figure 6.8. Reimagining Yonge Street. Source: Yonge Redevelopment Plan (1968)

above it, would make a significant contribution to multilevel urbanism (see figure 6.9). The pedestrian network is divided into major and minor routes, as well as major and minor nodes where these new streets would meet. In addition to these nodes, there would be a number of above-ground walkways over the proposed ring road (Beecroft and Doris Avenues). The major pedestrian network would lead south and cross over Highway 401 at two points.

Willowdale's redevelopment was to follow a linear growth pattern with development concentrated along major and minor nodes on Yonge Street. Given the intensity of car traffic, pedestrian traffic at these nodes would be "discouraged as much as possible," with the bulk of pedestrians directed to either underground passages or above-grade crossings, as shown on the concept plan (Jones & Parkin, 1968b, p. 79). In terms of design, "graphics, lighting and street furniture" were to be coordinated with this "sequence and rhythm of spatial development" (Jones & Parkin, 1968b, p. 77). At these nodes, the building forms would intensify. This pattern was to suit the rhythms of both car drivers and pedestrians: "for the automobile user the rhythm of the street space opening and closing identifies, at the speed of the automobile, the intensity and nature of the uses proposed, while the pedestrian is oriented to Yonge

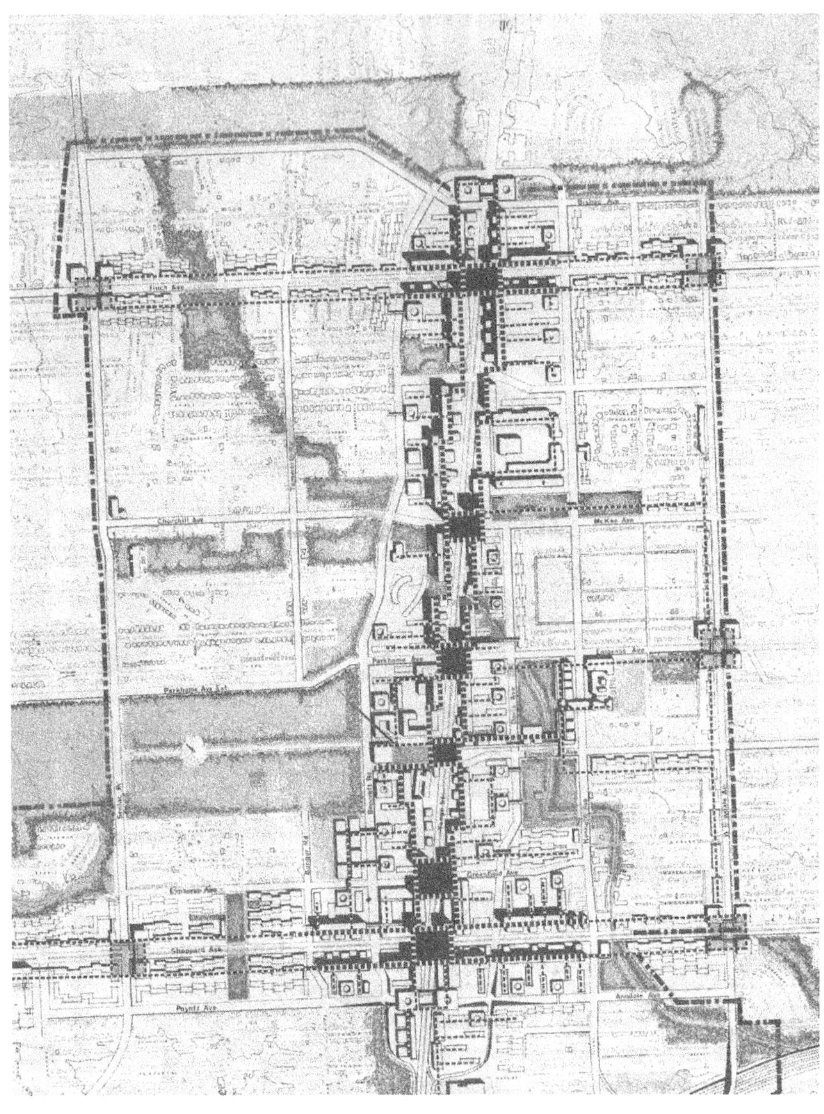

Figure 6.9. Detail of the pedestrian-traffic nodes along Yonge Street. Source: Yonge Redevelopment Plan (1968)

Figure 6.10. Rendering of the Civic Square farmers' market situated at the base of the tower. Source: Yonge Redevelopment Plan (1968)

Street only at points of intense common activity" – that is, the three central nodes: Sheppard, North York Centre, and Finch, which are the three subway stops in the centre (Jones & Parkin, 1968b, p. 103). The most intense node would be that of the Civic Square (at today's North York Centre), which would include both sides of Yonge Street, joined by City Hall. On the west side of Yonge Street would be the civic square, a 1,000-foot communications tower complete with revolving restaurant, and an above-ground parking lot with a permanent farmer's market underneath at ground level and whose "colourful stalls and fresh produce," Mayor Service claimed in a 1968 *Globe and Mail* article, would "give the extensively planned new sub-city … a needed touch of abandon, disorder and surprise." On the east side of Yonge Street would be a performing arts centre, an "Art Complex," and a gymnasium with underground parking.

Linked to these nodes would be a major street-level, north–south pedestrian system grade on both the east and west sides of Yonge Street; this would be the "focus for all local activities" (Jones & Parkin, 1968b, p. 80). Local retail, low-rise housing, and other neighbourhood facilities would be situated between Yonge Street and the pedestrian walkway; it would face the walkway and also offer a buffer zone between the high-rise centre and the existing suburban areas outside the development area. The extensive pedestrian system would act as a ring around the entire development extending further south to York Mills Station and also offering a way to cross Highway 401.[7] At points, the walkway would be covered, either by an arcade or a completely enclosed mall. It is unclear from the plan if Jones and Parkin knew how extensive the coverage would be (there are no drawings of these pedestrian spaces), although on the occasion of the plan's unveiling, Parkin said in a local report in *The Enterprise* newspaper that the enclosed pedestrian areas – walkways, malls, and overpasses – were typical of Canadian architecture and the winter climate. In this aspect of the plan, Parkin and Jones drew upon some of the existing outdoor pedestrian shopping streets, and in particular many of the shopping malls that Gruen featured in *Shopping Towns USA* (1960): in addition to Cumbernauld and Vällingby, they also cited Gruen's now demolished Midtown Plaza in Rochester, New York. Gruen did not just build the indoor shopping malls for which he is best known, but also outdoor public gathering spaces and squares. Like many architects and planners discussed in this book, Gruen's models were the Lijnbaan Shopping Centre in Rotterdam, designed by Team 10 architect J.B. Bakema, as well as the outdoor shopping centre in Vällingby, and the town centre of Harlow.

Much of the urban design in the plan coupled with the widening of surrounding roads can be read through this one goal: assuring the mobility of all, pedestrian and cars, in a well-ordered, hierarchical landscape. The many road-widening schemes proposed in the plan were easily dismissed as outside the budget and little consideration was later given to any sort of pedestrian network. The utopian aspects of the plan lie in the belief that parallel cities could so easily be constructed, as envisioned in the plan's separations. This was not simply about building some pedestrian overpasses or underground pedestrian walkways, but also offering an interconnected link of enclosed and open pedestrian spaces, including the pedestrian bridges over the highway, and the street-level pedestrian areas parallel to Yonge Street.

The monumental gesture of a city hall straddling Yonge Street was complemented by an equally monumental communications tower (this was before it had been decided that the CN Tower would occupy its current downtown location). Here the tower – proposed to be anywhere from 600 to 1,300 feet in height – with its revolving restaurant would loom over Yonge Street, a monument to the new scale of transportation and communication to which Jones and Parkin alluded.[8] In their effort to build a core around a central monument, they echoed city builders of the past who sought to erect monuments in the places they believed a core would develop. In *The Heart of the City*, Giedion echoes this attention to history: "In Rome, Sixtus V had the imagination and foresight to place his obelisks on spots where he felt a Core could arise, and around them some of the most beautiful squares of Rome have since developed" ("Conversation at CIAM 8," 1952, p. 39). In a 3 May 1969 *Toronto Daily Star* article entitled "North York: Where a New High-rise Changes the Skyline Once a Week," Parkin said:

> People need a distinctive group of buildings for their sense of loyalty. Toronto has the City Hall, London has Westminster Abbey, Paris has the Arc and the Eiffel Tower and so on. The role of architecture is a profound one in this loyalty and North York's problem is an absence of a specific symbol people can attach themselves to. I'm looking out the window of the classic example of what I'm talking about: Los Angeles, 60 suburbs in search of a city.

In many ways, Parkin's remark echoes Carver's own plea in *Cities in the Suburbs*, where he asks, "Can we leave nothing permanent behind"? (1962a, p. 75). He was referring to the monumental architecture of, say, a church or a town hall, an architecture that sought to combat what Carver called the forces of "ubiquitous mobility" and "anti-nucleation." Parkin's reference to the exploding landscape of Los Angeles reiterates this view: Willowdale, with its communications tower and megastructure-like city hall, would offer a counter-image to the forces of suburbanization.

The communications tower and the city hall, rising above and around the street, were to be Willowdale's monumental symbols of the changes in transportation and communication that Jones and Parkin believed made both the city and the suburb obsolete. One could imagine that if it was built, the tower would not mourn the death of God like Georges Bataille's obelisk, but rather the death of the street and a human scale that, Blumenfeld claimed, had been eclipsed by the extra-human scale of cars, freeways, skyscrapers. In this way, the tower and the city hall stand for the logic of modernism's majesty and authority, and the removal of the sidewalks are a symbol of the invasive and almost brutal approach to the existing urban landscape.

Implementation, or Burn the Model

In February 1969, Jones and Parkin's plan was debated by North York

Council, but due to protests, largely from ratepayers' associations in the area, its approval was delayed.[9] (People carried placards reading "Burn the Model" and "Nuts to Hi Rise Living.") One month later, over 600 residents attended another council meeting specifically to discuss the plan; they were overwhelmingly in opposition to the "high-rise" plan, as it was referred to in a *Toronto Daily Star* article, particularly the plan they believed would lead to the demolition of 1,200 homes over a 20-year period to make way for the high-density development.[10]

The development area Parkin and Jones proposed would go ahead as planned: high-rise residential and office buildings would line the east and west sides of Yonge Street, with concentrations around the subway stops. The development would extend as far as Beecroft Avenue to the west and Doris Avenue to the east. As time went on, however, many of the plan's monumental gestures were left on the drawing table: the civic centre straddling Yonge Street, the removal of pedestrians from Yonge Street, the tower, and the pedestrian malls on either side of the street, as well as the many street widenings and extensions that the authors had planned. By 1969, Parkin had moved to Los Angeles to take up practice, and both he and Murray Jones were no longer involved with the redevelopment.

In many ways the story of Willowdale's implementation offers some striking parallels with South City's: even though South City's plan paid particular attention to the details of the living environment, the varied architecture, the living streets, and the multifunctional city centre, there simply was neither the funds nor the time to develop it; all energy was directed to building apartments as fast as possible. The problem in Willowdale was not lack of funds or resolve, but concerted opposition to the monumental gestures of the plan and general public opposition to high-rise building in a suburban environment. The inhabitants lamented the end of the suburbs, while the architects and planners set the stage for it.

The period following the Willowdale plan's release coincides with, and indeed is inseparable from, the revolt against the heroic brand of modernist urbanism that began with Jane Jacobs and *The Death and Life of Great American Cities*. Whereas the implementation of the South City plan stemmed in part from the prioritizing of mass production over the architecture of the socialist environment as a whole, the turn away from modernist urbanism in Toronto, and Willowdale in particular, meant the rejection of some of its core architectural tenets: monumentality, the separation of cars and pedestrians, and the outright privileging of motor vehicle traffic and road building. Jones and Parkin prepared their plan amidst the intense conflict around expressway development in the city that began with the 1943 Master Plan, which in addition to the Toronto Bypass (now Highway 401) included the inner-city Spadina Expressway and the Crosstown Expressway, both of which were canceled in 1971 in the face of downtown protest (see Sewell, 1993, pp. 177–82). Thus, the rejection of the monumental details of Jones and Parkin's plan, along with the road widenings and sidewalk removals, should be read as part of the wider rejection of "expressway modernism" and the practices of separating pedestrians and cars.

Mayor Service was voted out of office the following year by residents who were unhappy about the changes to their suburban environment, and so it was Service, rather than North York's celebrated Mayor Mel Lastman, elected in 1972, who bore the ire of suburban residents unhappy with the new development. Although much of Willowdale was redeveloped in the 1980s under Lastman, it was Service who commissioned the first plan and bore the brunt of the initial wave of outrage and protest from citizen's groups. Although Service himself negotiated some of the initial deals for the land that would become part of the centre, his ambition made him an architect's mayor, someone who strongly believed in the architectural visions of the day. Lastman, on the other hand, was a businessman who brought development and developers to Willowdale.

When North York planners revisited Willowdale's redevelopment plans in the early 1970s, now under the title of "The Yonge Street Centre Area," they deemed Jones and Parkin's plan "too grandiose"; they wanted instead to avoid "monumentality" (Matthew & Davidson, 1983, p. 1). As a result, they wrote that the scale of the plan was "simply unacceptable and provoked concerted and aggressive public opposition" (Matthew & Davidson, 1983, p. 1).

When a new redevelopment plan was being formulated in 1977, extensive public consultations took place. The turn away from the heroic modernist principles of separation is illustrated by the submissions from the Ward 9 South Residents' Association, led by a young Jack Layton.[11] Although many submissions were made at these consultations from a number of individuals, businesses, and resident associations in the area, Layton's group most succinctly summed up the contentions within modernism, particularly in the reform movements in Toronto in the 1970s (the group was not against the development, although they were concerned about the heights of the proposed buildings). In its proposal Layton called for subsidized housing in the new centre, but he argued that this should "not be concentrated in particular buildings, where the stigma of a ghetto can become attached to it." He also wrote that "we are greatly concerned about the pedestrian orientation of the new downtown area," particularly the tendency to isolate pedestrians from the "urban experience by enclosing them in malls or walkways within building structures." Layton also maintained the "right of non-motorized private vehicles to be in the downtown area" and argued for bike lanes on Sheppard and Finch Avenues and Yonge Street. Planners needed to find a way to link the three nodes to "encourage strolling" and maintaining the centre's pedestrian orientation. He also emphasized preserving Willowdale's "grid street pattern" and "not allowing developments to take over and close streets." The preservation of this pattern will "limit the size of developments" (Yonge Street Centre Area Plan, 1979, n.p.).

Many of these points were part of the larger reconsideration of modernism that was also going on in 1960s Prague. Like Jiří Lasovský, Jones and Parkin were part of a long line of modernists who argued for separating cars and pedestrians through the creation of networks of pathways separated from the neighbourhood roads. Yet Lasovský's liveable streets were not simply a design principle, they were part of the larger project of humanizing both modernism and socialism across the sídliště, not just in a few isolated centres. Jones and Parkin's neglect of the existing street life on Yonge Street suggests a dehumanized modernism in favour of automobile circulation.

Since the 1980s, 60 new condominiums have been built in the redevelopment area along Yonge Street with office buildings at the key nodes of the Sheppard and North York Centre subway stations (see figure 6.11). This necessitated the removal of many of the street's historical buildings and the two-storey brick structures that housed a diverse range of independent businesses (Blackett, 2009, p. 44). It is the densest neighbourhood in Toronto. The North York Civic Centre was eventually built in 1979 and an adjoining North York Public Library was built by noted Toronto architect Raymond Moriyama in 1985, along with a hotel and a multilevel shopping arcade (shown in figures 6.12 and 6.13). The square surrounded by these buildings has been developed over the years and includes an artificial stream, benches, and trees, along with a bandshell and an outdoor skating rink. Although it is situated next to seven lanes of traffic on Yonge Street, *Spacing* magazine co-founder Matthew Blackett calls it "one of the most dynamic public spaces outside the city core" (2009, p. 53).

Although the monumental modernism of Jones and Parkin's plan was rejected, Layton's recommendations – especially his calls to preserve the area's grid street pattern – went unheeded. Yet one key aspect of Jones and Parkin's plan was for the most part kept: because of the potential for residential traffic mixing with the increased car traffic that was going to come with redevelopment, the authors proposed creating a ring road to contain the redevelopment and separate it from the

Figure 6.11. The competing built forms of this post-suburban landscape: condominiums and the low-rise fabric of the post-war period. Source: Steven Logan

Figure 6.12. North York City Centre, which includes Mel Lastman Square (centre), with hotel, businesses, shops, central library, and subway access (right) and the offices of the Toronto District School Board (left). Source: Steven Logan

Figure 6.13. View inside the pedestrian arcade as part of the North York Civic Centre (with escalators to the subway and underground shops). Source: Steven Logan

suburbs. Construction work did not begin on the ring road until the mid-1980s, but its effects can be felt today: it not only diverts traffic from Yonge Street, but functions as a physical barrier separating the high- and low-density developments, the general feature of the plan which is most visible today. The ring road is one of most important and lasting elements of Willowdale's modernist urbanism; indeed, its final design and route was only approved in 1985, with construction only recently finished.

Figure 6.14. Separating the suburb from the city on Willowdale's ring road: bollards and a cul-de-sac. Source: Steven Logan

The ring road has two main functions: to carry traffic as efficiently as possible between Sheppard and Finch Avenues, and to separate the new suburban downtown from the traditional suburbs beyond. But the ring road is more than a simple design gesture, and like the crane in South City, it brings together the interests of traffic engineers, architects, and urban planners with asphalt, cul-de-sacs, and traffic bollards (see figure 6.14). As such, it is a key marker in Willowdale's landscape of circulation, one that simultaneously allows and restricts the mobility of both cars and pedestrians. By acting as a barrier to the quiet suburban roads, it helps preserve the dominant suburban way of life.

And yet the discussions around the ring road and especially its scale – which is particularly evident when we consider what exists on the other side of it – illustrate how both intense development and single-family homes could be accommodated in what may easily be termed an example of residential reterritorialization. The "road network layout" was key to this, with the ring road serving as the medium through which the goal of reterritorialization could be achieved: "the impossibility of people passing through," as Eric Charmes puts it (2010, p. 357). It is a leftover of the modernist planning of Blumenfeld and

Figure 6.15. Looking north from one of the key nodes: Sheppard Avenue and Yonge Street. Source: Steven Logan

others who sought to unite the subcentre through the separation of cars and pedestrians, city and suburb.

The ring road creates a superblock in this suburban downtown, albeit one fundamentally different from the superblock of Stein and Wright's Radburn. The superblock is the downtown, and through the ring road and the network of cul-de-sacs, its aim is to separate the city from the suburb. Although planned, its concentration of condominiums and skyscrapers makes it far more dense than any of the pre-existing superblocks. And unlike Carver's vision, the edge city superblock, rather than separating work, recreation, and dwelling, brings them together. Parkin and Jones attempted to remake the suburban setting by creating an environment of and for automobility and urbanity that would also open the space to massive development. And yet the ring road also attempts to preserve a suburban way of life at the same time that it allows for urban growth. It is a landscape of sub/urbanity.

Conclusion

Not long after he was elected in 1965, Service promised in a speech on the planned future of North York that "by the year 1975, the City of North York shall be … the envy of the free world and the world wide Mecca of Planners." By 1975, however, North York was still arguing over how to develop its downtown. In 1974, a new plan had been accepted. In an article in the local North York paper, *The Mirror*, on 3 July 1974, the plan's author, John Bonnick, in a seeming reference to the earlier plan, said, "[the planners] do not want a concrete oasis at all – we should try to keep away from the monumentality of any buildings in the area and strive for a people-oriented complex."

In 1979 the Borough of North York officially became a city, but it still did not have the downtown its politicians wished for. Echoing Service's own 10-year grand claims, Lastman proclaimed in 1985 that "we'll have the most beautiful, most modern downtown in the world. Just give us another 10 years." Lastman had just asked Moriyama and Teshima Architects (1985) to prepare a "Streetscape Progamme" for the stretch of Yonge Street between Sheppard and Finch Avenues, which he submitted in 1985, and later revised (and significantly toned down) in 1990. Like the attempts to humanize architecture in Prague in the 1970s and '80s (in the "late beautiful" and postmodern phase), Moriyama calls for a "people place" that renews the spirit of the strong pedestrian spaces that were part of Jones and Parkin's original redevelopment plan, but where the previous plan had called for the removal of all pedestrians from Yonge Street, Moriyama sought the exact opposite: to make Yonge Street a pedestrian-oriented thoroughfare complete with a "Heritage train," a "Yonge Street 'Radial' tram" that would run up and down Yonge Street, thereby evoking the turn-of-the-century Radial train; as such, it would be part of an overall strategy to highlight the heritage of Yonge Street, which would include making buildings like the Gibson House "more accessible from Yonge Street." Reminiscent of Le Corbusier's plea in *Urbanisme*: "We must plant trees!" (Le Corbusier, 1924, p. 78), Moriyama's first suggestion in his proposals was to plant "a continuous boulevard of trees on each side of Yonge Street" to recall the "great urban streets of the world such as the Champs Elysees in Paris" (n.p.). Moriyama also revived Jones and Parkin's plan for tree-lined walkways that would "tie together the communities east and west of Yonge street." It was soon discovered, however, that the complex infrastructure under the sidewalks would not allow for tree planting, and so when Moriyama's proposals were revisited in 1990, trees would occupy a centre median down Yonge Street offering a reprieve to those trying to cross the busy seven lanes of traffic, while the sidewalks would have to make due mainly with concrete planters. Moriyama also proposed focusing activities on the sidewalk with "continuous street arcades" or canopies to protect people from inclement weather and "street level store front activities" that would reinforce the "people intensive street character" of Yonge Street. The street's branding would be completed by "Gateways" announcing the entrance to North York's downtown.

Some ten years later, Lastman admitted to John Sewell that "the streetscape turned out like hell. It's awful. It's not what I wanted" (1996, p. 17). Sewell wrote that the stretch of Yonge Street that runs through Willowdale, with its seven lanes of traffic, had become "a rushing river of noisy, dusty, smelly, dangerous vehicles" that is virtually impossible to cross, "nor is it fun walking on the sidewalk beside this mayhem" (1996, p. 17). Sewell writes that a downtown needs a street that "pulls things together," not spaces of separation, which often take the form of underground retail or above-ground walkways, both of which are "deadly for street life" (1996, p. 21). If South City was a product of the state-building apparatus, Willowdale was a product of the developers who, with their condominium and office towers, came to shape the Willowdale landscape.

The fate of Moriyama's ideas and many of the other modernist visions point to the tensions between an Athens Charter urbanism focused on circulation and a more humanized modernism of the kind advocated by Team 10 as well as socialist architects, focused on living environments and living streets. Here there were distinct parallels as socialist architects and urbanists continued to advocate for living streets into the 1970s and '80s, even if they were now disconnected from the wider program of socialism with a human face.

In a second report called "A Streetscape Vision for Downtown North York," Moriyama called North York's "instant downtown," along with the "conduit" for car traffic that is Yonge Street, a "psycho-social and planning contradiction that will take time and immense political will to overcome" (1990, p. 2). The aim is to make Yonge Street one of the "great streets": "Great streets achieve an integration of interests. They become the meeting place of individual and collective needs. They have places for lovers and places for festive events. A great street has a soul" (Moriyama, 1990, p. 3). In this way, we have returned to both CIAM's claims to reform Athens Charter urbanism, which influenced socialist architecture and urbanism in the 1960s, particularly in the plans for South City, and the calls for liveable streets in the sídliště.

For Robert Fishman (1987), suburbanization in the post-war period indicated not just another form of middle-class place privilege, but a new kind of city that he called a "technoburb" and that Joel Garreau (1991) would later call an edge city. For Fishman, this was not the culmination of suburbia, but its end. If the suburbs no longer exist, then Willowdale and South City were harbingers of the post-suburban edge city.

Conclusion: Unearthing the Suburban Core

Suburbs will not go away, nor should they. They may well hold the key to the solution of urban problems that were hitherto deemed insoluble.

– Robert Stern, 1981

There is a hill in South City that has become a favourite spot for residents, a place that in all of its earthiness is part of both the suburbs of history and the history of the suburb, even though it was never meant to *be* at all (see figure 7.1). Here is the architect Ivo Obrstein discussing the impacts on South City of fulfilling the state's building quotas: "They met the plans, but they took away the surrounding nature. They did not manage to take the earth from the building site, and so an artificial ground formed there … which separated the neighbourhood from the Milíčovský forest" (2016, pp. 55–6). The *halda*, as it is known (literally "heap" in Czech), was made up from the "leftover" dirt and stone removed when the ground was prepared to build South City, as well as bits of cable, brick, and the odd piece of concrete (Mikuláš, 2008).

The story of the halda begins with Lasovský's desire to use the leftover debris to construct a rolling landscape for South City's Central Park. Although the current park bears some remnants of this idea, with small rolling hills, the halda remained, much to the dismay of the architects. In the architects' view, the halda signalled everything wrong with crane urbanism: the destruction of nature, the disregard for the designs of the architects, and the unwillingness of the state to invest in the living environment of the sídliště. And yet in spite of the modernist attempts to create a work of landscape art, it became so on its own thanks to a decision in the early 1990s by a bureaucrat at IPS, a Czech state building and construction company, to plant all the trees that they had left over. As geologist Radek Mikuláš explains,

Figure 7.1. The milíčovská "halda" (heap) in South City. Source: Steven Logan

the company planted timber species that were cultivated from many different natural habitats, and "surprisingly everything grows so well, practically nothing has dried out or died" (2008, n.p.). It has created a veritable diversity of biological life. The landscape has become a refuge for residents, even a place to have a small campfire and spend the night.

The halda is as much a material marker as it is a symbol for the heap of ideas, both remembered and forgotten, that have come to define modernist urbanism and the unfulfilled visions of the modernist suburb. Ideas do not easily translate into built form, and the disparity between vision and built form defines the modern as much as that which was built. The cultural strata of urban space evolve like geological formations, albeit on a much different timescale. In a palimpsest, certain layers remain hidden from view. The *OED* describes a palimpsest as a "multi-layered record"; in geographical terms, it is "a structure characterized by superimposed features produced at two or more distinct periods." Its original meaning is tied to writing, as a "parchment or other writing surface on which the original text has been effaced or partially erased, and then overwritten by another."

As this book has reiterated at various points, the history of the suburbs is not simply a history of competing urban forms – single-family homes versus apartment blocks – but also a history of alternatives that included collective solutions over the privatism of single-family homeownership, from Ebenezer Howard's garden city to the varied communal spaces of the Scandinavian architects and urbanists who inspired their Czech counterparts. In many ways, Willowdale and South City represent two extremes: on the one hand, a developer-lead cosmopolitan downtown, the goal of which is density for density and profit's sake, and on the other, the increasingly privatized world of the sídliště and the new suburban spaces of a progressively isolated Czech Republic.

The layered histories of both South City and Willowdale mark them as emblematic places in the history of alternatives to the dominant suburbanisms of the post-war period: the cookie-cutter, mass-produced suburbs of North America and the prefabricated concrete apartment blocks on the socialist periphery. They are also key examples of the history of modernist urbanisms in the plural. Although Willowdale and South City look very different, it is too simplistic to suggest that these contrasting forms illustrate two distinct ideological spheres, the capitalist and the socialist. The suburban forms of the socialist city, particularly the plans for the city centre and the living streets in the 1960s, drew as much on the traditions of CIAM as they did on the socialist models of a classless society in a green city. To argue that it is one or the other is to simplify what was a complex process of negotiation that involved much more than the state merely dictating to architects and planners what a socialist settlement should look like. There was also a distinct socialist component to the living environment – the emphasis on the collective over the individual, regulated rent over homeownership, and the decommodification of land. Willowdale, although never making claims to borrow from the East, drew inspiration from a diverse range of sources that included Cumbernauld and Vällingby, and on a body of work on subcentres and cores by figures like Humphrey Carver, Jaqueline Tyrwhitt, and Hans Blumenfeld, all of whom came to Canada with international experience.

Although socialism is usually reviled for its authoritarianism, in this case, the ambitious plans for South City were more typical of new town planning in Scandinavia or the United Kingdom. In Willowdale the initial plans, with their desire to eliminate the sidewalks on Yonge Street, showed a complete disregard for street life that was authoritarian in its own right. This is the kind of thinking usually associated with modernism, not the living streets of the socialist suburbs. In both the capitalist and socialist (and now post-socialist) contexts, these places have struggled with their urbanity.

Sub/urbanity or Urbanity Denied?

The modernist urbanism of the 1960s envisioned urbanity without typical downtown streetscapes; however, urban commentators continue to favour the downtown streetscape as the marker of urbanity, and so perpetuate a city-suburb dichotomy. Urbanity does exist on the periphery, planned and unplanned.

Outside of the city centres, the street was not always the centre of attention because the street was often not conducive to social gathering. Social life happened in backyards, along rivers, or at the intersections around streetcar stops. In the suburbs, then, rather than forcing people to ride their bikes in a car-dominated landscape, we might reconsider the bias inherent in making the street itself the locus of change. This is not an apology for sprawl or automobility – on the contrary: it suggests that at its core, modernist urbanism sought to foster the growth of collective and public gathering spaces in the suburbs. For example, the socialist suburb's superblock, be it in South City or Poruba, allows children a degree of independence on the car-free green, and parents can still call from the apartment window to their children below, even if they are 10 floors up. Socialist urbanism assured, in theory, that everyone had equal access to a strong pedestrian realm, not just those who could afford it, and that those spaces were not simply the modern-day temples of consumption, shopping malls and hypermarkets.

Certainly, the sídliště have long been perceived as lacking in urbanity, and not only since the fall of socialism, when it became easier to criticize and vilify these places. In a 1990 speech in Prague's Old Town Square, Václav Havel said that one of the best things about Prague's metro is that "every day, tens of thousands of people from South City can go to another city, where they can see a real city, and not just the strangeness in which they live" (Havel, 1990). Havel had a particular antipathy for South City, preferring a different kind of urban form: "Every main street will have at least two bakeries, two sweet-shops, two pubs, and many other small shops, all privately owned and independent." The inhabitant, he argued, would again be able to "experience the phenomenon of home" in "family houses, villas, townhouses and low-rise apartment buildings" rather than the "totalitarian era" apartment blocks (quoted in Hirt, 2012, p. 34).

Ironically, a similar argument was made by socialist architects, like Lasovský in his plans for South City: it would be a place people would not just inhabit but call home. In South City, the human scale was given political meaning through "socialism with a human face": the living environment was not just a design gesture, but an attempt to reform both socialism and the sídliště. Given the political situation in the 1970s, the kind of public engagement enjoyed in places like Willowdale was not possible, but Lasovský continued to advocate for living streets, low-rise structures, and an active street-level environment. Lasovský's critiques would later be echoed in the post-socialist critiques of the sídliště as unfinished and in need of revitalization. South City's status on the periphery – not only on the edge of Prague itself, but as itself a place without a centre – is a sign of its sub/urbanity. The sídliště are, as one definition of suburban, somehow less than urban. Musil calls it the "torso-like character" of many of Prague's districts, of which South City is only the most remarkable case. Vítězslava Rothbauerová writes that the sídliště and its concrete apartment blocks will "never became a classical city, which grows and develops gradually over hundreds of years" (Rothbauerová, 2009, n.p.). Unlike Havel, Rothbauerová is not so quick to dismiss the sídliště; indeed, she points out that the sídliště also have many advantages, like attractive green spaces and easy access to the countryside. The absence of social and public spaces is not simply a feature of the sídliště, says Rothbauerová, but also of the existing suburbs further out, where the inhabitants of single-family houses need a car to buy a loaf of bread and take the children to school: for all of their deficiencies, at least the sídliště planned for shops and other facilities that people could walk to. There are no such plans in Prague's post-socialist suburbs. And as Hirt (2012) points out, these areas are not filled with independent pubs, bakeries, and sweet shops, but are instead dominated by corporate-owned hypermarkets like Tesco.[1]

Rostislav Švácha similarly argues that suggesting the sídliště should look more like a city centre is both an "inadequate and incorrect" approach because from its very inception the sídliště was built according to a set of ideas that are very different from those informing medieval cities like Prague or Olomouc (2000, n.p), even though it was, rightly or wrongly, compared with these older cities. Dolores Hayden (1984) reminds us that if the street is the privileged form of downtown life, this is not so in suburbs, where the centre of urban life, historically, has been the communal courtyard of the superblock, where a different kind of social life might prevail, not necessarily the one defined by cafes, boutique shops, and the bars of urban street life. For better or for worse, in the majority of urban spaces in the sídliště, the street is not the social centre of attention.

The fate of South City's city centre is interesting in this regard. Although the proposed site remains a grassy field, and the local aboveground centres are neglected, both aesthetically and socially, people can always shop in the surrounding hypermarkets or at one of the city's largest shopping malls, located on the west side of the D1 highway, in the South City II development.[2] The area has turned into a kind of edge city, a "regional commercial centre" (Sýkora & Mulíček, 2014, p. 152) that includes a corporate office cluster called the Park, complete with manicured lawns and glass facades. Its 120,000 square metres of office space is occupied by tenants such as IBM, 3M, Dell, Honeywell, Samsung, and Sony. In this way, it has taken on edge city qualities in the post-socialist period, although interestingly enough, it is on the site of what was to be South City's industrial area, which, if it had been built, would have provided nearly 16,000 jobs. South City's urbanity, then, becomes increasingly associated with the edge city urbanity of so many suburban spaces around the world, even as it retains its socialist suburban qualities.

Willowdale of course did get its centre – that was the plan all along. And for all of John Sewell's vitriol against modernism in *The Shape of the City* – Willowdale is very much a product of modernist urbanism – he writes that in Willowdale "the opportunities for a successful suburban downtown seem apparent" (1993, p. 220). At the same time, critics of Willowdale's rampant condo development have longed focus on its lack of urbanity and street-level urbanism, particularly on the part of urban design, as the conclusion to the preceding chapter showed. Toronto writer and former *Spacing* editor Shawn Micallef claims in a 2018 *Toronto Star* article that "the very 'downtownness' of it [Willowdale] is being denied." In 2018 and again in 2019, the city denied funding for yet another street-level program – this one entitled Transform Yonge and that would have brought bike lanes to Yonge Street and reduced car traffic from three lanes in each direction to two, affirming Jack Layton's calls for bike lanes back in the 1970s, when the city first conducted public consultations on the area. In the wake of the criticism of Athens Charter urbanism in Toronto and the cancelling of urban expressways like the Spadina Expressway, Ontario premier Bill Davis claimed in 1971 that, "If we are building a transportation system to serve the automobile, the Spadina Expressway would be a good place to start. But if we are building a transportation system to serve people, the Spadina Expressway is a good place to stop." Ostensibly this meant starting with bike and transit infrastructure. And yet, throughout the 1970s and '80s, this became an empty, if not altogether vague, promise, particularly when it came to bike infrastructure.

Going back to Moriyama's plans, Willowdale's lack of urbanity has been consistently associated with the amount of space devoted to cars. Urban theorist Pierre Filion shows that 43% of the land in the redevelopment area is devoted to cars, as opposed to 26% in downtown Toronto (2001, p. 151). Although many of Jones and Parkin's automobile-oriented plans were not realized, the area is still overrun by cars, even with the presence of a subway. Another target for criticism is the multileveled urbanism. Although modernist urbanism is critiqued for its attempt to put pedestrians on either underground walkways or above-ground flyovers, modernist thinking was also about street-level urbanism, which can be seen even in the work of Victor Gruen. In an article criticizing Willowdale's streetscape, John Sewell (1996) writes: "the last thing any downtown needs is a pedestrian bridge taking pedestrians off the sidewalk. Pedestrian walkways in the air are as deadly for street life as underground malls." New developments like the two towers that now dwarf the Gibson House museum, itself the reconstructed residence of Willowdale's most noted pioneer, tout the fact that residents would have direct access to the subway from their building: they do not have to go outside to leave their apartments. Mimicking the interconnected downtown of Calgary's +15 above- and below-ground walkway system, a similar urban dystopia presents itself: you do not need to go outside to accomplish everyday tasks; apartments are connected to the underground system, which itself is connected to a shopping mall, and which leads to the subway, taking you right to your downtown office.

Willowdale has become Blumenfeld's city in motion: "not only do people, goods, and messages move in the city, but the city itself is in motion. Day in and day out, structures are erected, altered and demolished" (1967b, p. 12). Much of the criticisms of Willowdale's lack of urbanity has to do with the demolishing of the two-storey buildings that used to line Yonge Street, many of which were independent businesses, as well as the massive increase in traffic on Yonge Street's seven lanes. John Filion, the long-time public servant, has seen most of the changes in Willowdale up close, first as an elected official in North York beginning in 1982, and then as a city councillor for Willowdale since 1991. In his foreword to the Leona Drive Project catalogue, Filion waxes nostalgic over the days when "people really knew their neighbours" – neighbours they could meet in the local bakery: "I have been trying for seventeen years now to get a good bakery in the area." Willowdale is not without its independent businesses, of course, and there is a tinge of ethnocentrism to Filion's claim, as some of the most interesting spaces of the Willowdale downtown are the small Korean karaoke bars, restaurants, and cafes that occupy the old two-story fabric, which is dwarfed by the condominiums and generic banks, coffee shops, and big box stores.

Filion's comment does put him in good company, including Havel. In the global suburb, Havel and Filion's bakery as a social meeting place, as a core, has become an anachronism, gone the way of the white picket fence or reborn as the artisanal bakery in the gentrified city. This is not to say good bakeries are unimportant to neighbourhoods – that is beside the point. Their parallel comments illustrate the challenges of urbanizing suburbs, a challenge that modernist architects and urbanists took up collectively in the 1960s. The idea that the suburban conversation was strictly a North American one is simply false: many cities struggled with built-up peripheral areas; opening the discussion to the socialist suburb changes what gets counted as a suburb and brings much-needed diversity to a debate that pits city against suburb. This places Willowdale and North York at the forefront of global suburbanisms, not simply because of the area's now radically diverse population, but because of its importance to the debates on the need to rethink modernism. Filion was Transform Yonge's biggest supporter, and so he was an important part of addressing the legacy of car-dominated planning. Familiar figures stalk Toronto's struggle with modernist urbanism – Clarence Stein, Ebenezer Howard, Le Corbusier, Frank Lloyd Wright. Undoubtedly, these figures are important; this is especially the case for Howard, whose reach extended to Czechoslovakia. But the comparison presented here with South City opens up the debate to include the modernists who criticized modernism, like Jaqueline Tyrwhitt in the 1960s, thereby adding to the breadth of influences on Toronto's suburbs. The phenomenal growth in North York is so important to Toronto as a whole, and yet in the suburban histories Don Mills is usually accorded the most importance.

In the post-Fordist era it is to Willowdale rather than Don Mills that we should turn to take the pulse of suburbanization. Willowdale fulfills one of Joel Garreau's stipulations for an edge city, one that Micallef points out in his article: "from farmers' field or village to a dense urban core in just half a century or so." Toronto is the only Canadian city featured in Garreau's list of edge cities – Willowdale's centre is one of four that he cites.[3] Willowdale should be seen alongside other established edge cities like Reston, which was also built on the principles of modernist urbanism, and whose original settlement also preceded its modern development.

Willowdale's "uptown downtown" has a clear edge defined by its ring road, and within this edge city superblock it has become "a regional end destination for mixed use" (Garreau, 1991, p. 7), including for jobs, residences, entertainment, shopping, and its many bars and restaurants. Willowdale is typical of what Garreau calls "Uptowns," by which he means "Edge Cities built on top of pre-automobile settlements" (1991, p. 113), with their traditional urban grid landscape of sidewalks and street-facing shops – this was the very landscape that the Skelly Plan and Jones and Parkin's plan sought to do away with completely. But, as Garreau notes, uptowns are "surrounded by entrenched neighbourhoods." As opposed to new edge cities, they have histories of urbanization and "layers of development" and are usually the first to get transit systems. This is one of the things that makes Willowdale as an edge city palimpsest so interesting, as between the rows of condominiums – its latest and most dominant layer – one can read the other historical layers of the edge city. If Willowdale is a microcosm of late twentieth-century post-suburbia, a "sequel to the suburbs" (Phelps, 2015), it is here, in the tangle of condominium towers and single-family houses, that the alternatives to the suburb are emerging.

Whereas the superblock was designed to keep traffic out, Willowdale's development seeks to keep the traffic *in*, and off of the residential streets beyond: a superblock turned inside-out. The result is that Yonge Street continues to be dominated by traffic congestion while the surrounding streets have been sealed off by cul-de-sacs, bollards, and other street-altering mechanisms. It was an effort to preserve the city-suburb divide within Willowdale. "We are selling bungalows in the sky," claimed a marketer for the Avondale subdivision, at the south end of Willowdale, "at prices that are less than they would be for teardowns in the immediate neighbourhood." The "teardowns" refer to the small CMHC-designed houses that formed the core of the early post-war suburb and which are fast disappearing from the landscape, deemed too small for today's families, with much larger houses built in their place (sometimes referred to as "monster" homes). They may have been the proper setting for family life in the post-war period, but in Willowdale today family life is just as often found in the condominium towers – the bungalows in the sky – as it is in the single-family homes.

Although new Willowdale residents are assured mobility, it is not always welcome: signs in front of new developments warn prospective buyers that their children may not have access to the local schools. Here is a sign of the death of the neighbourhood unit in the spaced-out scale of this urban subcentre. Every weekday morning one can see lineups of children along the ring road waiting for a bus that will take them to schools outside the district. A former school trustee for the area, Mari Rutka, described the situation as a "'vertical city,' a virtual town in the air without schools," in contrast to the "flat city" (Brown, 2012) across the ring road, where neighbourhoods were deliberately built around a school. Calling to mind the streets in the air of Le Corbusier's Unite d'habitation residential building in Marseilles, Rutka says, "we need a school in the air, but right now that idea's up in the air – it's too expensive" (Brown, 2012).

The fate of the neighbourhood unit and the single-family house as the most desired form of accommodation resonates not only in Willowdale, but across the spectrum of global suburbanization. The neighbourhood unit has become increasingly irrelevant because it only applies to areas of low-density, single-family homes. And here we find one of the drawbacks of the suburban city centre model, particularly as envisioned by Carver, who sought to update the neighbourhood unit with his cities in the suburbs while at the same time maintaining a rigid dichotomy between single-family homes and apartment buildings: a family raised its children in a house, not an apartment, which was for retired couples, individuals, or childless couples. The struggles of Willowdale's high-rise families go back to these very problems. The subheading for a 2018 *Toronto Star* article on the topic reads, "While highrises shot up along the Yonge St. corridor in North York, it was as if nobody predicted – or planned for – the day they would eventually be full of children." In the original plan, the average size of a household in the apartment buildings and condominiums that now line the street was to be 2.2 occupants per household. Willowdale's development boundary magnifies the contrast between two deeply divided forms of suburban living: high-rise and single-family homes. The kids still cannot walk to school.

Developments like South City remind us that single-family homes do not have a monopoly over suburban dwellings, and that public spaces and collective life could, in theory at least, be at the centre of suburban life. In South City, the socialist approach to dwelling, which focused on collective rather than private life, assured that families would occupy apartments. Much of the renewed interest in the Czech sídliště, their history and their revitalization, addresses both the individual apartments and the collective spaces. If the sídliště were derided in the immediate post-1989 period as soulless, a view epitomized by Havel's antipathy for the sídliště as synonymous with totalitarianism, then the renewed interest in the diverse and varied architectural and dwelling histories of the sídliště suggest a parallel to the renewed interest in the periphery in the face of increasingly gentrified city centres. The living environment of the sídliště should not only be seen as "utopia made real" (Zadražilová, 2013) in the sense of the architectural avant-garde, but also as the socialist attempt to create an alternative to the capitalist suburb, the kind of single-family house suburbs and edge city developments that in the post-socialist period have developed beyond the borders of South City in places like Pruhonice, Čestlice, and Nupaky, as well as the shopping mall and edge city developments in South City II-West.

If we consider together the criticisms of and praise for Willowdale and South City, both in the past and in the present, it becomes increasingly difficult to call them either a suburb or a city. And with good reason. Parkin and Jones specifically set out with the idea that the terms "city" and "suburb" had become obsolete, and the sídliště in the post-war period offered a completely new living environment that residents were hard-pressed to describe it as a city. Jones and Parkin claim that the city and the "downtownness" that Micallef appeals to was connected to an economy of manufacturing and forms of transportation and communication that were no longer dominant in 1967. The compact urban area was not an appealing design principle, but one necessitated by, and developed in tandem with, streetcars. The separation of work and home that developed alongside new transportation and communication technologies – cars, elevators, subways, and telephones – contributed to an urban explosion. As individual cities gave way to metropolitan regions, Jones and Parkin deemed both the city and the suburb obsolete. It was in this vein that they turned to the idea of urban subcentres to describe places like Willowdale, while admitting that there was little precedent for this in 1967. Here was a clear attempt to rethink the suburbs on both a planning and a conceptual level by offering a prevision of the coming edge city and technoburb, which has taken on renewed importance in the twenty-first century to make suburbs more resilient to climate change and less automobile-oriented. Roger Keil's concept of sub/urbanity accounts for the diverse ways in which the living environments of places like Willowdale and South City are at once urban and suburban.

Jones and Parkin's vision was so thoroughly imbued with automobility that it is unsustainable in the twenty-first century. If architects, urbanists, planners, and others are to offer not totalizing visions, but gestures towards a philosophy beyond the city-suburb dichotomy, we would do well to look back into the history of the suburbs before moving forward. First, the privatism of suburban life has been shown to not just be a feature of single-family homes, but also the sídliště of the 1970s and '80s. Single-family homes themselves are not the problem, but rather "corporate profiteering" and "a desire for privatism" fostered not only by "consumer capitalism" (Keil, 2018, p. 76) but also by consumer socialism.

In one sense, the suburbs, as a singular space somehow separate from the city, have gone away. Obviously, they have not disappeared, but in a time of global urbanization and technological interconnection, it is hard to see any built-up space as somehow separate, no matter how high the gate or how intolerant the residents. And yet the diverse histories of the suburbs, not necessarily visible in the built landscape, may indeed "hold the key to the solution of urban problems that were hitherto deemed insoluble" if urban theorists, planners, and dwellers alike are able to move past the unhelpful "city good, suburb bad" distinction and towards a global urban space rich in possibilities. Ideas, once derided, have again become popular: they are no longer just dwellers in the suburbs of history, but now very much in the present, calling for trees and green spaces in the city, minimum-sized and affordable apartments, urban gardening, communal dwellings, common greens, trams and streetcars. Being together means finding common ground between the people who live downtown, those who live in the inner suburbs, and the people on the fringes, and also between past, present, and future. The goal is to ultimately find common ground on the ground beneath our feet. In the search for new visions of the urban periphery, we could do worse than to learn from the suburbs of history and the histories of modernist urbanism in Willowdale and South City.

Notes

1. Introduction: Crossing Divides

1 Keil (2018) cites Sewell's (1993, 2009) work on Toronto and for the United States, James Kunstler (1993). In the socialist context, the housing estates in their designs are more often than not associated with the Athens Charter.

2 As urban planner and theorist Patsy Healey (2013) argues in her work on "The Transnational Flow of Planning Ideas and Practices," "images of urban form, promoted through international networks of architects, and movements such as CIAM … flowed vigorously around the world" and they in turn influenced the modernization projects of the politicians and planners in power (1512).

3 Hirt (2012, 2017). See also the collection of essays in Siegelbaum (2011) on "Mobility and Socialist Cities," which use Le Corbusier as a reference point. Elke Beyer (2011) in that collection points to the diverse set of ideas that influenced socialist architects and urbanists in their envisioning of city centres in the GDR and the USSR.

4 See especially *Team 10 East: Revisionist Architecture in Real Existing Modernism*, edited by Łukasz Stanek and Dirk van den Heuvel (2014). Marcela Hanáčková's contribution looks specifically at the Czech context.

5 One exception is Gordon's (2018) recent article on Carver's contribution to community planning, which I discuss in chapter 5.

6 Metropolitan Toronto was formed in 1953 as a federation made up of 13 municipalities, including North York. In this two-tiered level of government, "Metro" took on regional responsibilities, while each of the municipalities oversaw local issues. In 1967, this was reduced to six municipalities, and in 1998 Metro was done away with altogether with amalgamation of all the municipalities into the City of Toronto.

7 According to Sýkora and Mulíček (2014), nearly one-third of Prague residents had a country house; the country houses made up for people's

inability to own property or land in the city, and also allowed them to escape from the confines of their drab environs (137). Historian Paulina Bren (2002) argues that this private ownership was tolerated by the Czechoslovak regime because it kept people out of the cities, where they might cause political trouble, and in their homes. It was in line with the stated goals of the so-called normalization period of the 1970s and '80s: the "quiet life" and the "policy of peace" (Bren, 2002, 123).

8 In the third volume of the three-volume publication *East West Central: Re-Building Europe, 1950–1990*, Maroš Krivý (2017) addresses the critique of the sídliště in Czechoslovakia in the 1970s and '80s, a critique which had already begun in the 1960s with the discourse on humanizing architecture.

9 These three basic ideas of Freund and Martin's seminal text *The Ecology of the Automobile* (1993) are reworked and developed in John Urry and Mimi Sheller's (2000) work on the system of automobility and also in *Against Automobility* (Böhm et al., 2006)

10 Although there have been recent publications exploring socialist consumerism and leisure (e.g., Crowley & Reid 2010), leisure and free time was an important topic for Czech sociologists in the 1960s, particularly when it came to distinguishing socialist free time, influenced by Marxist humanism, from capitalist leisure. Like their capitalist counterparts, Czech sociologists, like Blanka Filipcová, attempted to address the increasing leisure time of the masses (see Miljački, 2017, pp. 239–67).

11 This comment should remind Torontonians of Rob Ford's claim that he ended the war on cars by eliminating a car registration fee in the first days of his tumultuous mayoral tenure in 2010.

12 In the Czech Republic and Slovakia, South City is certainly not peripheral to discussions of urban history. Although rarely, if ever, discussed as a suburb, it is one of the most well-known sídliště, along with Petrzalka, which was built around the same time in Bratislava, and thanks to Václav Havel's particular antipathy for it and the sídliště in general.

13 See the voluminous output associated with the Global Suburbanisms project (http://suburbs.info.yorku.ca).

2. Looking for the Antithesis of the Suburb

1 I want to thank one of the anonymous reviewers for pointing me towards the *Plan of Chicago* as one of the first comprehensive modern plans to focus on separating cars and pedestrians.

2 Stein was a consultant on the initial urban plan for Chandigarh, developed by Albert Mayer, a colleague of Stein's at the RPAA. The crux of Mayer's plan was the "superblock or neighbourhood unit" (Avermaete & Casciato, 2014, p. 136). The design of the superblocks fell to Polish architect Maciej

Nowicki, who grouped the residences around courtyards, and whose designs made an impression on the Indian authorities. However, his tragic death in a plane crash not long after was the beginning of the end of Mayer's plan. Shortly after, a new team was brought in, led by Le Corbusier.

3 See also Schuyler (1986).
4 Cycling is rarely, if ever, discussed in the literature on the modern city. It is as if the bicycle, as a mode of transportation, and cycling simply vanished with mass automobile use, at least in the modern visions of the city (Furness, 2010).
5 Quoted by Mar Ermers in *Der Wiener Tag*, 18 October 1931
6 Interestingly enough, Canada was one of three non-European countries also considered as hosts for the congress, which reflected the belief that CIAM's ideas were now, in the post-war period, a worldwide concern.
7 Havlíček in a letter to fellow architect Bohuslav Fuchs in 1956 (Hanáčková, 2014, p. 75).

3. Socialist Space

1 The question is addressed most directly by Andrusz et al.'s (1996) edited collection on the socialist city and also taken up by Sonia Hirt (2012) in her monograph on Sofia. See also, more recently, Zarecor (2018).
2 The most recent scholarship on Teige comes in the form of a massive compendium translated into English (Michalová, 2018). Peter Zusi (2004, 2013) has also written on Teige's versions of constructivism and poetism and his connection to figures like Walter Benjamin.
3 Teige in a letter to Sigfried Giedion, 17 May 1930 (quoted in Spechtenhauser & Weiss, 1999, p. 223).
4 Spechtenhauser and Weiss (1999) note that there is no documentary evidence to confirm if the lectures on housing and the city actually took place. Hannes Meyer, with whom Teige had close ties, was dismissed on 1 August 1930 (Spechtenhauser and Weiss, 1999, p. 237). An article based on the lecture was published in the journal *ReD* (*Revue Devětsil*) under the title "Towards a sociology of architecture" (Teige, [1930] 1977).
5 For work in English on Baťa's Czechoslovak origins, see Ševeček and Jemelka (2013) and Klingan and Gust (2009).
6 The landlord's rent is the interest paid on the initial amount used to buy the land for the garden city, which was always undeveloped agricultural land and thus more affordable (and which over time would inevitably increase if the city was built); the sinking fund refers to the principle amount paid, while the rates are the most relevant for the garden city's collective life, as they would go towards maintaining the municipal and

collective services. Eventually, the first two rents will be paid off, so that after a period of time all the rent will go towards maintaining the collective services, even though the rent paid continues to be the same.

7 For an in-depth discussion of the differences between the garden city as a social, utopian ideal and the way it was actually taken up in planning, see Fishman (1977, pp. 23–88).

8 Marx's *Manuscripts* is notable in large part for the way it circulated. Buck-Morss (2000) writes that it was "unknown and unpublished" until 1927, when it was released by the Marx Engels Institute in Moscow. This is the edition Teige was likely working with. The German translation did not appear until 1932, after which it became an important influence on Western Marxists (Buck-Morss, 2000, p. 118).

9 In socialist Czechoslovakia, when children received their red scarves as pioneers, they would recite the phrase *vždy připraven!* (always ready!).

10 See Guzik (2009) for an excellent overview of the collective dwelling from the perspective of gender studies.

11 In the years immediately following the Communist Party takeover of the Czechoslovak government in February 1948, Teige's ideas would come under attack. A 1947 essay entitled "Modern Architecture in Czechoslovakia," written at the urging of Sigfried Giedion and published in French and English, was heavily criticized in both English- and Czech-language Communist journals in 1948 following the Communist coup. In 1950, the Communist Party began a press campaign discrediting Teige (Dačeva, 1999, p. 381). In October 1951, a short while after this campaign, Teige died of a heart attack while waiting for the tram (Dluhosch, 2002, p. xi). His apartment was sealed by the police and most of his personal papers and library were taken, never to be seen again. His death was followed by the suicides of his two long-time female companions, Josefina Nevařilová (on the same day) and, ten days later, Eva Ebertová, another long-time female companion (Aulický, 1999, p. 386).

12 See Guzik (2009, pp. 408–9) and Miljački (2017, pp. 59–82) for a discussion of the Club for Reasonable Consumption and the idea of "Necessism." The group included Karel Honzík and Ladislav Žák.

13 The descriptions of Poruba and Solidarita's history that follow are indebted to Kimberly Elman Zarecor's work in *Manufacturing a Socialist Modernity* (2011, especially 150–76 and 268–71), as well as Herbert Guzik's work on Solidarita (2009, 2013). For further work on Solidarita, see Špičáková (2014).

14 This would become magnified, for example, in Zlín, the Baťa company town largely built along garden city principles in the 1920s during the capitalist period of the first republic. This is how the architect of Zlín's communal apartment building (1946), Jiří Voženílek, describes the

situation: "Capitalist leadership at Baťa supports unilaterally low-rise family buildings, because it wants to maintain the illusion that workers have a privately owned property and restrict the collective organization of workers outside the factory" (quoted in Guzik, 2013, p. 35).

15 The Panelací project, a reference to the prefabricated concrete panels used to construct housing, has thoroughly documented the sídliště throughout the Czech Republic, and presented their findings on a detailed website (panelaci.cz) and a two-part publication in Czech (Skřivánková et al., 2016, 2017), as well as one English publication (Skřivánková, Svácha et al., 2017).

4. South City as a Work of Art in the Age of Mass-Produced Dwellings

1 Interview with Jiří Musil, Prague, 3 July 2012. The interview was conducted in English.

2 Jiří Voženílek, the first head of this department, which was created in 1961, was one of the few members of the Czechoslovak interwar avant-garde who continued to be active in the post-war period and was very much influenced by Karel Teige's functionalism. One of his most significant contributions in the interwar period was his work on a master plan for Zlín, a garden city built for and around the Baťa shoe factory in the 1920s. In 1948, he became the first director of Stavoprojekt, the state organization that replaced private architectural studios, and in 1951 he became the director of the newly established Výzkumny ústavu výstavby a architektury (VUVA, Institute of Architecture and Town Planning), through which he advocated the industrialization and typification of dwellings (Zarecor, 2011, pp. 23, 262). He was also a consultant on the Etarea project.

3 Interview with Jiří Lasovský, Prague, 24 June 2012. The interview was conducted in Czech.

4 Lasovský interview, 24 June 2012.

5 Throughout the socialist period, architects and urbanists focused their attention on developing pedestrian spaces. In 1983, urban planners Jiří Hrůza and Blahomír Borovička wrote that the sídliště needed "living street spaces" (175). Krivý (2017) locates these discussions in the late socialist period, however, they had already begun in the 1960s with Hrůza's crisis of the sídliště (Hrůza, 1967a).

6 "Schválení podrobného územního plánu Jižního Města na území Chodov a Háje" (Approval of the detailed urban plan for South City on the lands of Chodov and Haje), Records of the meeting minutes of the bureau, council and local authorities ÚNV, NVP, a HMP (1945–1994), Archival record 28 December 1968, Prague City Archives.

7 Lasovský interview, 24 June 2012.
8 Interview with Vítězslava Rothbauerová, Prague, 8 June 2012. The interview was conducted in Czech.
9 Lasovský interview, 24 June 2012.
10 Lasovský interview, 24 June 2012.
11 Musil interview, 3 July 2012.
12 The film was never released publicly in Prague (Horton, 2002).
13 During communism, the surface parking lots in South City were called "unfillable" because of the waiting lists to buy a new car (Kotek 2009). These days, there are plans to finally build the parking garages, but parking in South City, both in these garages and on the street, will no longer be free. One South City myth is that it is harder to find a parking spot there than in the medieval core of the city. The proposed parking fee and the new multilevel parking garages have outraged residents and local politicians affiliated with Prague 11's governing party (as of 2016), Hnuti pro Prahu 11 (Movement for Prague 11)
14 Lasovský interview, 24 June 2012.

5. Redesigning the Post-war Suburban Landscape

1 The name was changed to the Canadian Mortgage and Housing Corporation in 1979.
2 Aside from Carver's three main books – *Houses for Canadians* (1948), *Cities in the Suburbs* (1962a), and *The Compassionate Landscape* (1975) – and a number of published essays, this chapter draws on unpublished material from the Humphrey Carver Archive at the Canadian Centre for Architecture in Montreal.
3 Although the practice can be traced back to nineteenth-century reformers, the logic of CIAM's urbanism in particular was that the old city was sick and needed a cure, which would be administered by the urban planner as surgeon. The focus on the city as a circulatory system certainly helped in this regard, and in the appendix to Le Corbusier's *Urbanisme* (1924) there is a series of anatomical drawings of the body's circulatory systems. In the chapter on "Physic or Surgery," he argues that Paris does not need physicians, but surgeons.
4 At the end of his life, when Carver clearly would have been familiar with Regent Park's problems, he remained an unrepentant advocate of social engineering, refusing to engage with the complexities of social life beyond the nuclear family and writing that there were many families without fathers, so "no wonder Regent Park became known as chaotic and a ghetto. To shelter and protect people with problems was the object of the exercise" (1994, p. 49). An exercise for some (architects), a life for others

(residents). Instead of questioning his own normative values around family life, he suggested that critiques of public housing had to do with their excellent design, which made them "stand out from the general dullness of the surrounding city" and thus put their low-income tenants into an unwanted spotlight (Carver, 1975, p. 142).

5 Although historian Richard White suggests that the neighbourhood unit remained largely unknown in Toronto until the planner Eugene Faludi began to experiment with it in the 1940s, Carver, while working in Toronto in the 1930s, had clearly already been influenced by Perry's thinking – Faludi approached Carver to co-write a book on urban planning and housing, but nothing came of it, which Carver attributed to their *sizeable* differences: at 6′6″, Carver towered over the 5′ Faludi (Carver, 1975, p. 59).

6 The 160-acre size was not arbitrary. Under the 1862 Homestead Act in the United States, any settler willing to pay a filing fee and live on and cultivate the land for five years was given 160 acres for free (Anderson, 2011, p. 120). I thank one of the anonymous reviewers for pointing this out. The roots of the individualism of home and land ownership, as Carver points out, stretch back to its rural origins.

7 For overviews of the founding and building of Don Mills, see Sewell (1993, pp. 79–96) and White (2016, pp. 103–13)

8 The letter was not unsolicited. It arose out of the possibility that CMHC would offer financial support for an accompanying publication.

9 The Massey Report makes no reference to the heritage of Canada's Indigenous population, which, given the time, is not surprising.

10 Carver's attention to the disappearing farmland in the area around Toronto was prescient. In 1970 the Toronto-Centred Region Plan called attention to the "quantities of land ... removed prematurely from agricultural and recreational use" within the "commuting area surrounding Metropolitan Toronto" (quoted in Sewell, 2009, p. 43). One of the areas along the corridor he was discussing, Prince Edward County, today offers well-heeled urbanites both a recreational and gastronomical experience – an authentic, and yet still commodified, experience of nature.

6. The "Total Image": The Making of Willowdale Modern

1 Metropolitan Toronto was a federation of municipalities formed in 1954 that included the City of Toronto and its 12 surrounding suburbs.

2 There are currently three subway stops that serve the area: Sheppard, North York Centre, and Finch. The subway was extended to Sheppard and Finch in 1974, with the North York Centre station added in 1987. North York Centre is situated right in the middle of what was the village of Willowdale. Although opened in 1987, transport planning consultant

Edward J. Levy notes that North York was already pressuring for the stop in the early 1970s while Finch station was under construction (2013, p. 102).

3 He had even enlisted filmmaker Christopher Chapman – who experimented with multiple screens in his film for Expo 67 – to make a film about North York (unfortunately, North York's councillors were not supportive and the film was never made).

4 Blumenfeld is a central figure in the circulation of people and ideas between East and West. Born in Germany, he lived and worked in the Soviet Union between 1930 and 1937, where he worked with the State Institute for Projecting Cities ("Giprogor"), first in Moscow then in Nizhni Novgorod, and where he became part of the cadre of international experts that descended on the Soviet Union in the 1920s and early '30s. See chapter 6 ("In the Soviet Union 1930–1937") of Blumenfeld's autobiography *Life Begins at 65* (1987). He soon emigrated to the United States and then moved to Toronto to work on Metropolitan Toronto's first regional plan in the 1950s.

5 Lynch (1960) and Appleyard et al. (1964).

6 Although Murray V. Jones and Associates and John B. Parkin and Associates were the firms commissioned, Murray Jones and John C. Parkin's signatures appear on the plan. Still, the plan was likely a collective project involving many actors, as there were over 200 employees at the Parkin firm alone (Fraser et al., 2013, p. 45), but for simplicity's sake, I will refer to the authors as Parkin and Jones. This is also complicated by the fact that there was another John Parkin at the firm, John B. (no relation to John C.), with whom John C. is frequently confused in newspaper articles of the time.

7 It is unclear whether Jones and Parkin assessed the feasibility of a pedestrian cycling route through the Toronto Don Valley Golf Course, built in 1956, on the west side of Yonge Street along the Don River.

8 For an overview of the ill-fated attempts to put a tower in Willowdale, see Osbaldeston (2008).

9 A ratepayers' association is an association of homeowners in a particular neighbourhood that deals with problems and issues that arise there. It is the equivalent of American homeowners' associations.

10 There was a wave of anti-high-rise sentiment in Toronto in the late 1960s and early 1970s, particularly around public housing projects in the city like Trefann Court and St. Jamestown, but the protest in Willowdale was different, largely symbolic, as there was no question that with the arrival of the subway development would proceed.

11 Jack Layton, then an aspiring politician, would go on to lead the New Democratic Party to one of its best results federally in 2011. He passed away later that year after a battle with cancer.

Conclusion: Unearthing the Suburban Core

1 Hirt (2012, p. 55) suggests that Havel's ideal is closer to the New Urbanist communities, which reproduce the Old World charm of streetcar suburbs and mythical small-town America.

2 In the original plans dating back to the 1980s, the architects had proposed a community centre on this site.

3 Garreau also lists Mississauga, which is a suburban municipality to the west of Toronto, "The Don Valley Parkway-401 area" just to the north of Don Mills, and curiously Yorkville, which in most people's eyes would be seen as part of the gentrified urban core, and not an edge city in any regard.

References

Adorno, T., & Horkheimer, M. (1947) 2002. *The dialectic of enlightenment: Philosophical fragments* (G. Schmid Noerr, Ed.; E.F.N. Jephcott, Trans.). Stanford University Press.

Anderson, H.L. (2011). That settles it: The debate and consequences of the Homestead Act of 1862. *The History Teacher, 45*(1), 117–37. Retrieved from https://eric.ed.gov/?id=EJ963318

Andrusz, G.D., Harloe, M., & Szelényi, I. (Eds.). (1996). *Cities after socialism: Urban and regional change and conflict in post-socialist societies*. Blackwell.

Appleyard, D., Lynch, K., & Myer, J.R. (1964). *The view from the road*. MIT Press.

Attributes of the core. (1952). In J. Tyrwhitt, J.L. Sert, & E. Rogers (Eds.), *CIAM 8: The heart of the city* (pp. 166–7). Lund Humphries.

Aulický, M. (1999). Postscript: My uncle, Karel Teige (Eric Dluhosch, Trans.). In E. Dluhosch & R. Švácha (Eds.), *Karel Teige: L'enfant terrible of the Czech avant-garde* (pp. 384–7). MIT Press.

Avermaete, T., & Casciato, M. (2014). *Casablanca/Chandigarh: A report on modernization*. Canadian Centre for Architecture, Park Books.

Baird, G. (1974). Architecture and politics: A polemical dispute, a critical introduction to Karel Teige's "Mundaneum," 1929 and Le Corbusier's "In defense of architecture," 1933. *Oppositions, 4*, 80–1.

Banham, R. (1976). *Megastructure: Urban futures of the recent past*. Harper & Row.

Bartoň, J. (1996). *Chodovská tvrz: K historii Prahy 11* [Chodov's citadel: A history of Prague 11]. Kreace.

Bartoň, J. (2014). *Z historie území dnešní Prahy 11* [A history of today's Prague 11]. Retrieved from https://www.praha11.cz/cs/praha-11-v-kostce/z-minulosti-prahy-11/z-historie-uzemi-dnesni-prahy-11.html

Benjamin, W. (1931) 1999. The destructive character (Edmund Jephcott, Trans.). In M.W. Jennings, H. Eiland, & G. Smith (Eds.), *Vol. 2 of Walter Benjamin: Selected writings* (pp. 541–2). Belknap Press of Harvard University.

Benjamin, W. (1933) 1999. Experience and poverty (Rodney Livingstone, Trans.). In M.W. Jennings, H. Eiland, & G. Smith (Eds.), *Vol. 2 of Walter Benjamin: Selected writings* (731–5). Belknap Press of Harvard University.
Berman, M. (1983). *All that is solid melts into air.* Verso.
Beyer, E. (2011). Planning for mobility: Designing city centres and new towns in the USSR and the GDR in the 1960s. In L.H. Siegelbaum (Ed.), *The socialist car: Automobility in the Eastern Bloc* (pp. 71–91). Cornell University Press.
Bittner, S. (1998). Green cities and orderly streets. *Journal of Urban History, 25*(1), 22–56. https://doi.org/10.1177/009614429802500102
Blackett, M. (2009, Summer–Fall). Downtown North York. *Spacing*, 52–3.
Blau, E. (1999). *The architecture of Red Vienna, 1919–1934.* MIT Press.
Blažek, B. (1998). Sídliště: Zrcadlo nastavené době [The sídliště: A mirror of our times]. *Umění a řemesla, 4*, 41–8.
Blumenfeld, H. (1967a). The role of design. *Journal of the American Institute of Planners, 33*(5), 304–10. https://doi.org/10.1080/01944366708977793
Blumenfeld, H. (1967b). Planning for the city in motion. *Habitat, 10*(1), 10–13.
Blumenfeld, H. (1987). *Life begins at 65: The not entirely candid autobiography of a drifter.* Harvest House.
Böhm, S., Jones, C., Paterson, M., & Land, C. (Eds.). (2006). *Against automobility.* Blackwell.
Bondy, E. (1983) 1992. *Cesta českem našich otců: Píkarský román-obžaloba* [The journey through our fathers' Czechia: A picaresque novel]. Česká expedice.
Borovička, B., & Hrůza, J. (1983). *Praha: 1000 let stavby města* [Prague: 1000 years of building the city]. Panorama.
Bren, P. (2002). Weekend getaways: The "Chata," the "Tramp" and the politics of private life in post-1968 Czechoslovakia. In D. Crowley & S.E. Reid (Eds.), *Socialist spaces: Sites of everyday life in the Eastern Bloc* (pp. 123–41). Berg.
Brown, L. (2012, February 6). Students in the lurch as condo crush forces school boundary changes. *Toronto Star.* Retrieved from https://www.thestar.com/news/gta/2012/02/06/students_in_the_lurch_as_condo_crush_forces_school_boundary_changes.html
Buck-Morss, S. (2000). *Dreamworld and catastrophe: The passing of mass utopia in East and West.* MIT Press.
Carver, H. (1935) 1975. *A housing programme* (The Research Committee of the League for Social Reconstruction, Ed.). University of Toronto Press.
Carver, H. (1939). *Planning Canadian towns* [Unpublished manuscript]. Folder 304, Humphrey Carver fonds,1932–1987, Canadian Centre for Architecture.
Carver, H. (1941). The strategy of town planning. *Journal RAIC, 18*(3), 35–40.
Carver, H. (1947). Planning for half a million houses. *Layout for Living, 2*, 1.
Carver, H. (1948). *Houses for Canadians.* University of Toronto Press.

Carver, H. (1957). Notes to senior staff course. File 257A, Humphrey Carver fonds, 1932–1987, Canadian Centre for Architecture.

Carver, H. (1961, October). Town centres: A proposal for centennial celebrations. *Listening Post*, pp. 3–4.

Carver, H. (1962a). *Cities in the suburbs*. University of Toronto Press.

Carver, H. (1962b). Housing: A search for focus. *Journal RAIC*, *39*(10), 59–66.

Carver, H. (1962c). H. Carver's *Cities in the suburbs* [Notes made in the course of writing the book, 1960–2]. File 303, Humphrey Carver fonds, 1932–1987, Canadian Centre for Architecture.

Carver. H. (1964). Letter to Herbert Hignett, president, CMHC (15 July). File 220, Humphrey Carver fonds, 1932–1987, Canadian Centre for Architecture.

Carver. H. (1965). A pride of cities [Preliminary notes for a book about Canadian cities]. File 257D, Humphrey Carver fonds, 1932–1987, Canadian Centre for Architecture.

Carver, H. (1966). Letter to W.J. Withrow, director, Art Gallery of Ontario (7 September). File 224, Humphrey Carver fonds, 1932–1987, Canadian Centre for Architecture.

Carver, H. (1967). New face of the city. File 257J, Humphrey Carver fonds, 1932–1987, Canadian Centre for Architecture.

Carver, H. (1968). Cities in the suburbs revisited [An address by H. Carver at the University of British Columbia]. File 257E, Humphrey Carver fonds, 1932–1987, Canadian Centre for Architecture.

Carver, H. (1975a). *Compassionate landscape*. University of Toronto Press.

Carver, H. (1975b). Speech to CHDC Awards Dinner. File 257J, Humphrey Carver fonds, 1932–1987, Canadian Centre for Architecture.

Carver, H. (1978a). The suburbs: Their purpose, growth and design. In *Suburbia: Costs, consequences, alternatives*. Proceedings of Symposium in the Urban Studies Programme, Spring 1977, York University.

Carver, H. (1978b). Building the suburbs: A planner's reflections. *City Magazine*, *3*(7), 40–5.

Carver H. (1979). The private and the social habitat [Paper given at Waterloo University in a series sponsored by the Canadian Housing Design Council]. File 132, Humphrey Carver fonds, 1932–1987, Canadian Centre for Architecture.

Carver, H. (1994). *Decades: A personal report on the past century*. Topcopy.

Čelechovský, G., Stehlík, J., & Sýkora, V. (1967). Etarea: Studie životního prostředí města [Etarea: Study of the living environment of the city]. *Architektura ČSR*, *13*(7), 399–408.

Charmes, E. (2010). Cul-de-sacs, superblocks and environmental areas as supports of residential territorialization. *Journal of Urban Design*, *15*(3), 357–74. https://doi.org/10.1080/13574809.2010.487811

Chatrný, J. (n.d.). *Lesná. Panelací.cz*. Retrieved 8 January 2018 from http://panelaci.cz/sídliště/jihomoravsky-kraj/brno-lesna

Clarke, N. (2012). Actually existing comparative urbanism: Imitation and cosmopolitanism in north-south interurban partnerships. *Urban Geography, 33*(6), 796–815. https://doi.org/10.2747/0272-3638.33.6.796

Club for New Prague. (1925). *Sedum přednášek o architektuře* [Seven lectures on architecture]. Klub architektů.

Cochrane, A. (2011). *Foreword to mobile urbanism: Cities and policymaking in the global age* (E. McCann & K. Ward, Eds., pp. ix–xi). University of Minnesota Press.

Cohen, J.-L. (2000). *Introduction to modern architecture in Czechoslovakia and other writings, by Karel Teige* (I.Ž. Murray & D. Britt, Eds., pp. 1–55). Getty Research Institute.

Colomina, B. (1994). *Privacy and publicity: Modern architecture as mass media*. MIT Press.

Conversation at CIAM 8. (1952). In J. Tyrwhitt, J.L. Sert, & E. Rogers (Eds.), *CIAM 8: The heart of the city* (pp. 36–40). Lund Humphries.

Cook, I.R. (2018). Suburban policy mobilities: Examining North American post-war engagements with Vällingby, Stockholm. *Geografiska Annaler: Series B, human geography, 100*(4), 343–58. https://doi.org/10.1080/04353684.2018.1428495

Cook, I.R., Ward, S.V., & Ward, K. (2014). A springtime journey to the Soviet Union: Postwar planning and policy mobilities through the iron curtain. *International Journal of Urban and Regional Research, 38*(3), 805–22. https://doi.org/10.1111/1468-2427.12133

Cooke, C. (1997). Beauty as a route to "the radiant future": Responses of Soviet architecture. *Journal of Design History, 10*(2), 137–60. https://doi.org/10.1093/jdh/10.2.137

Crowley, D., and Reid, S.E. 2010. *Pleasures in socialism: Leisure and luxury in the Eastern Bloc*. Northwestern University Press.

Dačeva, R. (1999). Appendix: Chronological overview. Translated by Eric Dluhosch. In E. Dluhosch & R. Švácha (Eds.), *Karel Teige: L'enfant terrible of the Czech avant-garde* (pp. 348–83). MIT Press.

Darroch, M. (2008). Bridging urban and media studies: Jaqueline Tyrwhitt and the explorations group, 1952–1957. *Canadian Journal of Communication, 33*(2), 147–69. https://doi.org/10.22230/cjc.2008v33n2a2026

David-Fox, M. (2006). Multiple modernities vs. neo-traditionalism: On recent debates in Russian and Soviet history. *Jahrbücher Für Geschichte Osteuropas, 54*(4), 535–55. https://www.jstor.org/stable/41051747

Deckker, T. (2000). *Introduction to the modern city revisited* (pp. 1–6). Spon Press.

Dluhosch, E. (2002). *Introduction to the minimum dwelling, by Karel Teige*. MIT Press.

Domhardt, K.S. (2012). The garden city idea in the CIAM discourse on urbanism: A path to comprehensive planning. *Planning Perspectives, 27*(2), 173–97. https://doi.org/10.1080/02665433.2012.646768

Dunnett, J. (2000). Le Corbusier and the city without streets. In T. Deckker (Ed.), *The modern city revisited* (pp. 56–79). Spon Press.

Enyedi, G. (1996). Urbanization under socialism. In G.D. Andrusz, M. Harloe, & I. Szelényi (Eds.), *Cities after socialism: Urban and regional change and conflict in post-socialist societies* (pp. 100–19). Blackwell.

Filion, P. (2001). Suburban mixed-use centres and urban dispersion: What difference do they make? *Environment and Planning A, 33*(1), 141–60. https://doi.org/10.1068/a3375

Fishman, R. (1977). *Urban utopias in the twentieth century: Ebenezer Howard, Frank Lloyd Wright and Le Corbusier.* MIT Press.

Fishman, R. (1987). *Bourgeois utopias: The rise and fall of suburbia.* Basic Books.

Forsyth, A., & Crewe, K. (2009). New visions for suburbia: Reassessing aesthetics and place-making in modernism, imageability and new urbanism. *Journal of Urban Design, 14*(4), 415–38. https://doi.org/10.1080/13574800903265470

Forty, A. (2012). *Concrete and culture: A material history.* Reaktion.

Fraser, L.M., McMordie, M., & Simmins, G. (2013). *John C. Parkin, archives and photography: Reflections on the practice and presentation of modern architecture.* University of Calgary Press.

French, R.A., & Ian Hamilton, F.E. (1979). *The socialist city: Spatial structure and urban policy.* Wiley.

Freund, P., & Martin, G. (1993). *The ecology of the automobile.* Black Rose Books.

Furness, Z. (2010). *One less car: The politics of automobility.* Temple University Press.

Gallagher, L. (2013). *The end of the suburbs: Where the American dream is moving.* Portfolio, Penguin.

Garreau, J. (1991). *Edge city: Life on the new frontier.* Doubleday.

Gehl, J. (1987). *Life between buildings: Using public space.* Van Nostrand Reinhold.

Giedion, S. (1941) 1969. *Space, time and architecture* (5th ed.). Harvard University Press.

Giedion, S. (1948) 1969. *Mechanization takes command.* W.W. Norton & Co.

Giedion, S. (1952a). Historical background to the core. In J. Tyrwhitt, J.L. Sert, & E. Rogers (Eds.), *CIAM 8: The heart of the city* (pp. 17–25). Lund Humphries.

Giedion, S. (1952b). The heart of the city: A summing-up. In J. Tyrwhitt, J.L. Sert, & E. Rogers (Eds.), *CIAM 8: The heart of the city* (pp. 159–63). Lund Humphries.

Giedion, S. (1958). *Architecture, you and me: The diary of a development.* Harvard University Press.

Giustino, C. M. (2012). Industrial design and the Czechoslovak pavilion at EXPO '58: Artistic autonomy, party control and cold war common ground. *Journal of Contemporary History*, *47*(1), 185–212. https://doi.org/10.1177/0022009411422371

Gold, J.R. (1997). *The experience of modernism: Modern architects and the future city, 1928–53*. E. & F.N. Spon.

Gold, J.R. (2006). The making of a megastructure: Architectural modernism, town planning and Cumbernauld's central area, 1955–75. *Planning Perspectives*, *21*(2), 109–31. https://doi.org/10.1080/02665430600555255

Gordon, D. (2018). Humphrey Carver and the federal government's postwar revival of Canadian community planning. *Urban History Review*, *46*(2), 71–84.

Gottlieb, M., & Todlová, M. (1969). Sociologické poznámky k jižnímu městě [Sociological comments on South City]. *Architektura ČSR*, *4*, 211–14.

Gronow, J., & Zhuravlev, S. (2010). Soviet luxuries from champagne to private cars. In D. Crowley & S. Reid (Eds.), *Pleasures in socialism: Leisure and luxury in the Eastern Bloc* (pp. 121–46). Northwestern University Press.

Gruen, V. (1964). *The heart of our cities: The urban crisis: Diagnosis and cure*. Simon & Schuster.

Gruen, V. (1960). *Shopping towns USA: The planning of shopping centers*. Reinhold.

Guzik, H. (2009). The Diogenes family: The collectivization of accommodation in Bohemia 1905–1948. Translated by Marta Filipová. *Art in Translation*, *1*(3), 381–417. https://doi.org/10.2752/175613109X12542203989165

Guzik, H. (2012). K prehistorii českých kolektivních domů [A prehistory of Czech collective housing]. *Gender rovné příležitosti výzkum*, *13*(1), 42–51.

Guzik, H. (2013). Kolektivní domy a sociální politika českých průmyslových podniků v letech 1939–1953 [Collective housing and the social politics of Czech industrial companies, 1939–1953]. *Umeni/Art*, *61*(1), 33–56.

Guzik, H. (2014). Hotelový dům [Residential hotel]. In L. Zikmund-Lender (Ed.), *Experimentalní Sídliště Invalidovna* [Invalidovna: An experimental sídliště] (pp. 62–73). Národní památkový ústav.

Hall, P. (1976) 2002. *Urban and regional planning*. Routledge.

Hanáčková, M. (2011). Dolní centrum Liberce v podání ateliéru sial: Tři přístupy při projektování center měst [Liberec Lower Town Centre by SIAL: Three approaches to designing town centres]. *Architektúra & Urbanizmus*, *45*(3–4), 200–19.

Hanáčková, M. (2014). Team 10 and Czechoslovakia: Secondary networks. In Ł. Stanek (Ed.), *Team 10 East: Revisionist architecture in real existing socialism* (pp. 73–100). Museum of Modern Art in Warsaw.

Harris, R. (2004). *Creeping conformity: How Canada became suburban, 1900–1960*. University of Toronto Press.

Hart, P. (1968). *Pioneering in North York: A history of the borough*. General Publishing Company.

Harvey, D. (1989). *The condition of postmodernity: An enquiry into the origins of cultural change*. Blackwell.

Hatherley, O. (2008). *Militant modernism*. O Books.

Hatherley, O. (2012). *A new kind of bleak: Journeys through urban Britain*. Verso.

Häussermann, H. (1996). From the socialist to the capitalist city: Experiences from Germany. In G.D. Andrusz, M. Harloe, & I. Szelényi (Eds.), *Cities after socialism: Urban and regional change and conflict in post-socialist societies* (pp. 214–31). Blackwell.

Havel, V. (1990). *Projev prezidenta ČSSR Václava Havla k výročí únorového převratu 1948* [Speech by President Vaclav Havel on the occasion of the 1948 Czechoslovak coup d'état]. Retrieved from http://www.vaclavhavel.cz

Hayden, D. (1984). *Redesigning the American dream: The future of housing, work, and family life*. W.W. Norton.

Hayden, D. (2009). *Building suburbia: Green fields and urban growth, 1820–2000*. Vintage.

Healey, P. (2013). Circuits of knowledge and techniques: The transnational flow of planning ideas and practices. *International Journal of Urban and Regional Research, 37*(5), 1510–26. https://doi.org/10.1111/1468-2427.12044

Henderson, S.R. (2013). *Building culture: Ernst May and the new Frankfurt initiative, 1926–1931*. Peter Lang Publishing.

Hess, P.M. (2005). Rediscovering the logic of garden apartments. *Places: Forum of design for the public realm, 17*(2), 30–5.

Hirt, S. (2012). *Iron curtains: Gates, suburbs, and privatization of space in the post-socialist city*. Wiley & Sons.

Hirt, S. (2017). O Sofia, where art thou? Suburbs as stories of time and space. In N.A. Phelps (Ed.), *Old Europe, new suburbanization? Governance, land, and infrastructure in European suburbanization* (pp. 66–84). University of Toronto Press.

Hirt, S., & Kovachev, A. (2015). Suburbia in three acts: The East European story. In P. Hamel & R. Keil (Eds.), *Suburban governance: A global view* (pp. 177–97). University of Toronto Press.

Honzík, K. (1963). *Ze života avantgardy* [From a life of the avant-garde]. Československý spisovatel.

Hopfengärtner, J. (2017). Introduction. In J. Hopfengärtner & Á. Moravánszky (Eds.), *Re-humanizing architecture: New forms of community, 1950–1970* (pp. 13–20). Birkhäuser Verlag GmbH.

Horowitz, G. (1966). Conservatism, liberalism, and socialism in Canada: An interpretation. *Canadian Journal of Economics and Political Science/Revue Canadienne d'Economique et de Science Politique, 32*(2), 143–71. https://doi.org/10.2307/139794

Horton, A.J. (2002). Against destruction: Věra Chytilová's Panelstory (Prefab story, 1979). *Kinoeye: New Perspectives on European Film, 2*(8). Retrieved from http://www.kinoeye.org/02/08/horton08.php

Howard, E. (1965). *Garden cities of tomorrow* (F.J. Osborn, Ed.). MIT Press. (First published in 1898 as *Tomorrow: A peaceful path to real reform.*)

Hrůza, J. (1962). *Budoucnost Měst.* Orbis.

Hrůza, J. (1967a). Krize sídliště [Crisis of the sídliště]. *Architektura ČSR, 13*(4), 1.

Hrůza, J. (1967b). *Město utopistů* [City of utopians]. Československý spisovatel.

Hrůza, J. (2006). Jiří Hrůza. Interview by Oldřich Ševčík and Lenka Popelová. In P. Urlich, P. Vorlík, B. Filsaková, K. Andrášiová, & L. Popelová (Eds.), *Šedesátá léta v architektuře očima pamětníků* [1960s architecture seen through the eyes of its observers]. Nakladatelství ČVUT.

Jacobs, J. (1958). Downtown is for people. *Fortune.* Retrieved from http://fortune.com/2011/09/18/downtown-is-for-people-fortune-classic-1958/

Jacobs, J. (1961). *The death and life of great American cities.* Random House.

Jíšová, K. (2005). Mraveniště lidí: Sny a realita budování socialistického velkoměsta [An ant-hill of people: Dreams and reality in the building of the socialist city]. *Dějiny a současnost: Kulturně a historická revue.* Retrieved from http://dejinyasoucasnost.cz/

Jižní město Pražské: Únorová diskuse u magnetofonu [Roundtable discussion on Prague's South City]. *Československý architekt, 16*(6–7), 1–3.

Keil, R. (2018). *Suburban planet: Making the world urban from the outside in.* Polity Press.

Ekers, M., Hamel, P. & Keil, R. (2015). Modalities of suburban governance in Canada. In P. Hamel & R. Keil (Eds.), *Suburban governance: A global view* (pp. 80–109). University of Toronto Press.

Kennedy, S. (2013). *Willowdale: Yesterday's farms, today's legacy.* Dundurn.

Khrushchev, N. (1963). On wide-scale introduction of industrial methods, improving the quality of and reducing the cost of construction. (Original speech delivered in 1954.) In T. Whitney (Ed.), *Khrushchev speaks: Selected speeches, articles, and press conferences, 1949–1961* (pp. 153–92). University of Michigan Press.

Kling, S. (2013). Wide boulevards, narrow visions: Burnham's street system and the Chicago Plan Commission, 1909–1930. *Journal of Planning History, 12*(3), 245–68. https://doi.org/10.1177/1538513213476709

Klingan, K., & Gust, K. (Eds.). (2009). *A utopia of modernity: Zlín: Revisiting Bata's functional city.* Jovis.

Kopp, A. (1970). *Town and revolution: Soviet architecture and city planning, 1917–1935* (T.E. Burton, Trans.). George Braziller.

Kotek, L. (2009). *Tady bylo Husákovo: Czechoslovakia 1982–1989* [Husák was here: Czechoslovakia 1982–1989]. Gema Art.

Kotkin, S. (1996). The search for the socialist city. *Russian History, 23*(1), 231–62. https://doi.org/10.1163/187633196X00150

Koukalová, Š. (2007). Hromadná bytová výstavba 70. let v severních Čechách [Mass housing in the 1970s in North Bohemia]. In L. Hubatová-Vacková & C. Říha (Eds.), *Husákova 3 + 1: Bytová kutura 70. let* [Husák's apartments: Housing culture in the 1970s] (vol. 3 + 1, pp. 75–92). VŠUP.

Krásný, J., Lasovský, J., & Řihošek, M. (1969). Podrobný užemn plán Jižního města v Praze [Detailed urban plan for South City in Prague]. *Architektura ČSSR, 7–8*, 442–48.

Krivý, M. (2016). Postmodernism or socialist realism? The architecture of housing estates in late socialist Czechoslovakia. *Journal of the Society of Architectural Historians, 75*(1), 74–101. https://doi.org/10.1525/jsah.2016.75.1.74

Krivý, M. (2017). Quality of life or life-in-truth? A late-socialist critique of housing estates in Czechoslovakia. In Á. Moravánszky, T. Lange, J. Hopfengärtner, & K.R. Kegler (Eds.), *Re-framing identities: Architecture's turn to history, 1970–1990: Vol. 3 of East west central: Re-building Europe, 1950–1990* (pp. 303–18). Birkhäuser.

Kroha, Jiří. (1935) 1969. Bytová otázka v SSSR [The housing question in Czechoslovakia]. In J. Cisařovský (Ed.), *Avantgardní architektura* (pp. 85–163). Československý Spisovatel.

Kunstler, J. (1993). *The geography of nowhere: The rise and decline of America's man-made landscape*. Simon & Schuster.

Lasovský, J. (1973). Jak roste město [How a city grows]. *Československý architekt, 2*, 1, 3.

Lasovský, J. (1975). Naděje a skutečnost [Hope and reality]. *Československý architekt, 15–16*, n.p.

Lasovský, J. (1981). Městský parter [City amenities]. *Architektura ČSSR, 6*, 248–55.

Le Corbusier. (1924). *Urbanisme*. G. Crès & Cie.

Le Corbusier. (1929) 1987. *The city of tomorrow and its planning* (F. Etchells, Trans.). Dover Publications.

Le Corbusier. 1973. *The Athens charter* (A. Eardley, Trans.). Grossman Publishers.

Le Corbusier. (1933) 1974. *In defense of architecture* (N. Bray, A. Lessard, A. Levitt, & G. B., Trans.). Reprinted in *Oppositions, 4*, 93–108.

Le Normand, B. (2011). Automobility in Yugoslavia between urban planner, market, and motorist: The case of Belgrade, 1945–1972. In L.H. Siegelbaum (Ed.), *The socialist car: Automobility in the Eastern Bloc* (pp. 92–104). Cornell University Press.

Lebow, K. (2013). *Unfinished utopia: Nowa Huta, Stalinism, and Polish society, 1949–56*. Cornell University Press.

Lefebvre, H. (1970) 2003. *The urban revolution*. University of Minnesota Press.
Lefebvre, H. (1974). *La production de l'espace*. Éditions Anthropos.
Lefebvre, H. (1974) 1991. *The production of space* (D. Nicholson-Smith, Trans.). Basil Blackwell.
Levy, E.J. (2013). *Rapid transit in Toronto: A century of plans, progress, politics and paralysis*. Neptis Foundation. Retrieved from www.levyrapidtransit.ca
Liscombe, R.W. (2007). Perceptions in the conception of the modernist urban environment: Canadian perspectives on the spatial theory of Jaqueline Tyrwhitt. In I.B. Whyte (Ed.), *Man-made future: Planning, education, and design in mid-20th century Britain* (pp. 78–98). Routledge.
Lizon, P. (1996). East Central Europe: The unhappy heritage of communist mass housing. *Journal of Architectural Education, 50*(2), 104–14. https://doi.org/10.1080/10464883.1996.10734709
Lynch, K. (1960). *The image of the city*. MIT Press.
Maier, K., Hexner, M., & Kibic, K. (1998). *Urban development of Prague: History and present issues*. ČVUT.
Masák, M. (2006). Mezi Expy [Between expos]. In P. Urlich, P. Vorlík, B. Filsaková, K. Andrášiová, L. & Popelová (Eds.), *Šedesátá léta v architektuře očima pamětníků* [1960s architecture through the eyes of its observers]. Nakladatelství ČVUT.
Matthew, M.R., & Davidson, A.B. (1983, April). *Review of the Yonge Street centre plan: Staff background report and proposed revisions*. North York Department of Planning and Development.
Maxim, J. (2009). Mass housing and collective experience: On the notion of microraion in Romania in the 1950s and 1960s. *Journal of Architecture, 14*(1), 7–26. https://doi.org/10.1080/13602360802705155
Maxim, J. (2011). The microrayon: The organization of mass housing ensembles, Bucharest, 1956–1967. In *Docomomo e-proceedings 4, postwar mass housing – East+ West*. Retrieved from https://sites.eca.ed.ac.uk/docomomoiscul/files/2012/11/p4_PAPER1_2.pdf
McCann, Larry. (2003). Suburbs of desire: The suburban landscape of Canadian cities, c. 1900–1950. In R. Harris and P. Larkham (Eds.), *Changing suburbs: Foundation, form and function* (pp. 111–45). Routledge.
McClelland, M. (2010). Modernist Architecture in North York. In *North York's modernist architecture revisited* (pp. 8–9). ERA Architects.
McFarlane, C. (2010). The comparative city: Knowledge, learning, urbanism. *International Journal of Urban and Regional Research, 34*(4), 725–42. https://doi.org/10.1111/j.1468-2427.2010.00917.x
Metropolitan Toronto Planning Board. (1970). *Metropolitan Toronto key facts*. Metropolitan Toronto Planning Board.
Metropolitan Toronto Planning Department. (1976). *Metroplan: Concept and objectives*. Municipality of Metropolitan Toronto.

Michalová, R. (2018). *Karel Teige: Captain of the avant-garde* (S. van Pohl, Trans.). Kant.

Mikuláš, R. (2008, April 19). Divočina na jižním městě [A wilderness in South City]. *Týdeník Respekt*. Retrieved from https://www.respekt.cz/tydenik/2008/17/divocina-na-jiznim-meste

Miljački, A. (2017). *The optimum imperative: Czech architecture for the socialist lifestyle, 1938–1968*. Routledge.

Moholy-Nagy, L. (1947) 1965. *Vision in motion*. Paul Theobold and Company.

Moravánszky, Á., Lange, T., Hopfengärtner, J., & Kegler, K. (Eds.). (2017). *East west central: Re-building Europe, 1950–1990* (3 vols.). Birkhäuser Verlag GmbH.

Moriyama & Teshima Architects. (1990). *A streetscape vision for downtown North York*. City of North York Planning Department.

Moriyama & Teshima, Planners Ltd. (1985). *Streetscape programme for Yonge Street centre area*.

MTARTS. (1967). *Choices for a growing region: A study of the emerging development pattern and its comparison with alternative concepts*. Ontario Community Planning Branch.

Mumford, E. (2000). *The CIAM discourse on urbanism, 1928–1960*. MIT Press.

Mumford, E. (2009). CIAM and the Communist Bloc, 1928–59. *Journal of Architecture, 14*(2), 237–54. https://doi.org/10.1080/13602360802704810

Mumford, L. (1938). *The culture of cities*. Harcourt, Brace.

Mumford, L. (1961). *The city in history*. Harvest Books.

Murray V. Jones and Associates Ltd. & John B. Parkin Associates. (1968a). *Yonge redevelopment study, Borough of North York*.

Murray V. Jones and Associates Ltd. & John B. Parkin Associates. (1968b). *Yonge redevelopment study*.

Musil, J. (1985). *Lidé a sídliště* [People and housing estates]. Nakladatelství Svoboda.

North York Planning Board. (1969). *District 11 plan*. North York Planning Board Board.

North York Planning Board. *Yonge Street centre area plan: Briefs, reports and minutes, January to April 1979*. North York Planning Board.

Novotný, J. (1978). Centrum pro 100,000 [City Centre for 100,000]. *Československý architekt, 15*, 4–5.

Nový, O. (1971). Introduction to Pražský projektový ústav [Introduction to Prague Design Institute]. In *Architekti Praze* [Architecture in Prague]. HDKN Praha.

Oberlander, P. (1998). Interview by Jim Donaldson. Peter Guo-hua Fu School of Architecture: Alumni interviews. Retrieved 16 December 2016 from https://www.mcgill.ca/architecture/people-0/aluminterviews/oberlander.

Obrstein, I. (2006). Interview by Lenka Popelová. In P. Urlich, P. Vorlík, B. Filsaková, K. Andrášiová, & L. Popelová (Eds.), *Šedesátá léta v architektuře očima pamětníků* [1960s architecture seen through the eyes of its observers]. Nakladatelství ČVUT.
Osbaldeston, M. (2008). *Unbuilt Toronto: A history of the city that might have been*. Dundurn.
Osborn, F. J. (1946) 1965. *Preface to garden cities of tomorrow by Ebenezer Howard* (F. J. Osborn, Ed., pp. 9–28). MIT Press.
Paperny, Vladimir. (2002). *Architecture in the Age of Stalin: Culture Two* (J. Hill & R. Barris, Trans.). Cambridge University Press.
Perry, C. (1929) 1974. The neighborhood unit. In *Neighborhood and community planning: Regional survey* (vol. 7). Regional Plan of New York and Its Environs. Reprint, Arno Press.
Perry, C. (1939). *Housing for the machine age*. Russell Sage Foundation.
Péteri, György. (2009). Streetcars of desire: Cars and automobilism in Communist Hungary (1958–70). *Social History, 34*(1), 1–28. https://doi.org/10.1080/03071020802628020
Phelps, N.A. (2015). *Sequel to suburbia: Glimpses of America's post-suburban future*. MIT Press.
Phelps, N.A., & Wu, F. (2011). *International perspectives on suburbanization: A post-suburban world?* Palgrave Macmillan.
Pucher, J. (1990). Capitalism, socialism, and urban transportation policies and travel behavior in the East and West. *Journal of the American Planning Association, 56*(3), 278–96. https://doi.org/10.1080/01944369008975773
Pucher, J. (1999). The transformation of urban transport in the Czech Republic, 1988–1998. *Transport Policy, 6*, 225–36. https://doi.org/10.1016/S0967-070X(99)00023-2
Řiha, C. (2007). V čem je panelák kamarád: Kvantitativní ohledy kvalit české panelové výstavby 70. Let [Our friend the panelák: A quantitive look at the quality of Czech panel building in the 1970s]. In L. Hubatová-Vacková & C. Říha (Eds.), *Husákova 3 + 1: Bytová kutura 70. Let* [Husák's apartments: Housing culture in the 1970s] (pp. 17–38). VŠUP.
Robinson, J. (2006). *Ordinary cities: Between modernity and development*. Routledge.
Rothbauerová, V. (2009). Sídliště [Microsoft Word file shared with the author].
Rubin, E. (2011). Understanding a car in the context of a system: Trabants, Marzahn and East German socialism. In L.H. Siegelbaum (Ed.), *The socialist car: Automobility in the Eastern Bloc* (pp. 124–40). Cornell University Press.
Schuyler, D. (1986). *The new urban landscape: The redefinition of city form in nineteenth-century America*. Johns Hopkins University Press.
Sequel to Etarea: Prague South Town. (1970). *Official Architecture and Planning, 33*(3), 230–2.

Sert, J.L. (1952). Centres of community life. In J. Tyrwhitt, J.L. Sert, & E. Rogers (Eds.), *CIAM 8: The heart of the city* (pp. 3–16). Lund Humphries.

Service, J.D. (1968, February 27). Remarks by Mayor Jas. D. Service on the occasion of the public presentation of the Yonge Redevelopment Plan at the Inn on the Park. North York Historical Society Scrapbooks, Book 22, 192–94.

Ševeček, O., & Jemelka, M. (Eds.). (2013). *Company towns of the Bata concern: History cases architecture.* Steiner.

Sewell, J. (1977). Where the suburbs came from. *City Magazine.* Special issue: John Sewell on the suburbs, *2*(6).

Sewell, J. (1993). *The shape of the city: Toronto struggles with modern planning.* University of Toronto Press.

Sewell, J. (1996, July 11–17). PSE of North York. *Now Magazine*, pp. 17, 22.

Sewell, J. (2009). *The shape of the suburbs: Understanding Toronto's sprawl.* University of Toronto Press.

Sheller, M., & Urry, J. (2000). The city and the car. *International Journal of Urban and Regional Research, 24*(4), 737–57. https://doi.org/10.1111/1468-2427.00276

Shoshkes, E. (2016). *Jaqueline Tyrwhitt: A transnational life in urban planning and design.* Routledge.

Siegelbaum, L.H. (2008). *Cars for comrades: The life of the Soviet automobile.* Cornell University Press.

Siegelbaum, L.H. (Ed.) (2011). *The socialist car: Automobility in the Eastern Bloc.* Cornell University Press.

Skřivánková, L., Švácha, R., Koukalová, M., & Novotná, E. (Eds.). (2017). *Paneláci 2: Historie sídlišť v českých zemích 1945–1989* [The paneláks 2: History of the sídliště in the Czech Republic, 1945–1989]. Museum of Decorative Arts in Prague.

Skřivánková, L., Svácha, R., & Lehkozivová, I. (Eds.). (2017). *The paneláks: Twenty-five housing estates in the Czech Republic.* Museum of Decorative Arts in Prague.

Skřivánková, L., Švácha, R., Novotná, E., & Jirkalová, K. (Eds.). (2016). *Paneláci 1: Padesát sídlišť v českých zemích* [The paneláks 1: Fifty sídliště in the Czech Republic]. Museum of Decorative Arts in Prague.

Spechtenhauser, K., & Weiss, D. (1999). Karel Teige and the CIAM: The history of a troubled relationship (Eric Dluhosch, Trans.). In E. Dluhosch & R. Švácha (Eds.), *Karel Teige: L'enfant terrible of the Czech avant-garde* (pp. 216–55). MIT Press.

Špičáková, B. (Ed.). (2014). *Sídliště Solidarita.* Texts in Czech by Michaela Janeckova, Eva Novotna, & Kimberly Zarecor. Archiv výtvarného umění.

Stanek, Ł. (2011). *Henri Lefebvre on space: Architecture,urban research, and the production of theory.* University of Minnesota Press.

Stanek, Ł., & van den Heuvel, D. (2014). Introduction: Team 10 East and several other useful fictions. In L. Stanek (Ed.), *Team 10 East: Revisionist architecture in real existing modernism* (pp. 11–33). Museum of Modern Art in Warsaw.

Stein, C.S. (1949). Radburn. *Town Planning Review; Liverpool, 20*(3), 219–63. Retrieved from https://www.jstor.org/stable/40101963

Stein, C.S. (1950) 1966. *Toward new towns for America*. MIT Press.

Švácha, R. (1995). *The architecture of New Prague 1895–1945* (A. Buchler, Trans.). MIT Press. (Originally published as *Od moderny kfunkcionalismu*. Odeon, 1985.)

Švácha, R. (2000). Ze sídliště neodejdu [I am not leaving the sídlište]. Interview by Petr Volf. *Reflex, 21*(50), 22–6. Retreived from http://www.jedinak.cz /stranky/txtŠvácha.html

Sýkora, L., & Mulíček, O. (2014). Prague: Urban growth and regional sprawl. In K. Stanilov & L. Sýkora (Eds.), *Confronting suburbanization: Patterns, processes, and management of urban decentralization in post-socialist Central and Eastern Europe* (pp. 133–162). John Wiley & Sons.

Sýkora, L., & Stanilov, K. (2014). Postsocialist suburbanization patterns and dynamics: A comparative perspective. In K. Stanilov & L. Sýkora (Eds.), *Confronting suburbanization: Patterns, processes, and management of urban decentralization in postsocialist Central and Eastern Europe* (pp. 256–295). John Wiley & Sons.

Szelényi, I. (1996). Cities under socialism and after. In G.D. Andrusz, M. Harloe, & I. Szelényi (Eds.), *Cities after socialism: Urban and regional change and conflict in post-socialist societies* (pp. 286–317). Blackwell.

Teige, K. (1923) 2000. Toward a new architecture. Reprinted in I.Z. Murray & D. Britt (Eds.), *Modern architecture in Czechoslovakia and other writings*. Getty Research Institute.

Teige, K. (1925–6). Obývací stroje [Machines for living]. *Stavba, 4*, 135–46.

Teige, K. (1927). *Stavba a báseň: Umění dnes a zítra, 1919–1927* [Building and poem: Art today and tomorrow, 1919–1927]. Vanek & Votava.

Teige, K. (1928) 2004. *Svět, který se směje* [A world that laughs]. *Vol. 1 of O humoru, clownech a dadaistech* (J. Thomáš, Ed.). Reprint, Akropolis.

Teige, K. (1929) 1974. Mundaneum. *Stavba, 7*, 145–55. Reprinted in *Oppositions, 4*, 83–91.

Teige, K. (1929–30) 2000. Ten years of the Bauhaus. In I.Z. Murray & D. Britt (Trans.), *Modern architecture in Czechoslovakia and other writings*. Getty Research Institute. Originally published as Deset let Bauhausu. *Stavba*, 8, 146–52.

Teige, K. (1930) 1977. K sociologii architektury [Towards a sociology of architecture]. In *RED: Měsíčník pro moderní kukturu* [RED: Journal of modern culture] (vol. 3, pp. 163–223). Jal-Reprint.

Teige, K. (1930) 2000. Modern architecture in Czechoslovakia. Reprinted in I.Z. Murray & D. Britt (Trans.), *Modern architecture in Czechoslovakia and other writings*. Getty Research Institute.

Teige, K. (1930) 2004). *Svět, který voní. Vol. 2 of O humoru, clownech a dadaistech* [The scent of the world. Vol. 2 of On humour, clowns and Dadaists] (J. Thomáš, Ed.). Reprint, Akropolis.

Teige, K. (1932) 2002. *The minimum dwelling* (E. Dluhosch, reprinted with preface and translated.). MIT Press.

Teige, K. (1933). *Práce Jaromíre Krejcara: Monografie staveb a projektů* [The works of Jaromír Krejcar: Buildings and projects]. Nakl.Václav Petr.

Teige, K. (1936) 1969. Vývoj sovětské architektury [The development of Soviet architecture]. In *Avantgardní architektura* [Avant-garde architecture] (pp. 9–163). Československý Spisovatel.

Teige, Karel. (1987). The housing problem of the subsistence level population: Summary of the national reports at the International Congress for New Building (CIAM), 1930 (C. Collins and M. Swenarton, Trans.). *Habitat International, 11*(3), 147–51. https://doi.org/10.1016/0197-3975(87)90025-7

Tyrwhitt, J. (1952). Cores within the urban constellation. In J. Tyrwhitt, J.L. Sert, & E. Rogers (Eds.), *CIAM 8: The heart of the city* (pp. 103–7). Lund Humphries.

Tyrwhitt, J. (1955) 1960. The moving eye. In E. Carpenter & M. McLuhan (Eds.), *Explorations in communication* (pp. 90–5). Beacon Press. Originally published in *Explorations, 4*, 115–19.

Tyrwhitt, J. (1962). In praise of exuberant diversity: "The death and life of great American cities" by Jane Jacobs. *Ekistics, 13*(77), 197–201. Retrieved from https://www.jstor.org/stable/43615982

Urlich, P., Vorlík, P., Filsaková, B., Andrášiová, K., & Popelová, L. (Eds.). (2006). *Šedesátá léta v architektuře očima pamětníků* [1960s architecture seen through the eyes of its observers]. Nakladatelství ČVUT.

Umbach, M. (2006). Urban history: What architecture does, historically speaking. *Journal of the Society of Architectural Historians, 65*(1), 14–15. https://doi.org/10.2307/25068230

Voženílek, J. (1967). Soutěž na South City v Praze [Competition for South city in Prague]. *Architektura ČSR, 2–3*, 91–9.

Wakeman, R. (2016). *Practicing utopia: An intellectual history of the new town movement.* University of Chicago Press.

Ward, S.V. (1999). The international diffusion of planning: A review and a Canadian case study. *International Planning Studies, 4*(1), 53–77. https://doi.org/10.1080/13563479908721726

Ward, S.V. (2016). *The peaceful path: Building garden cities and new towns.* Hertfordshire Press.

Weber, Nicholas Fox. (2008). *Le Corbusier: A life.* Alfred A. Knopf.

Welter, V.M. (2003). From "Locus Genii" to heart of the city: Embracing the spirit of the city. In I.B. Whyte (Ed.), *Modernism and the spirit of the city* (pp. 35–56). Routledge, Taylor & Francis.

White, R. (2016). *Planning Toronto: The planners, the plans, their legacies, 1940–80.* UBC Press.

Wigley, M. (2001). Network fever. *Grey Room, 4*, 82–122. Retrieved from http://www.jstor.org/stable/1262560

Williams, R. (1974). *Television: Technology and cultural form.* Collins.

Wilson, A. (1991). *The culture of nature: North American landscape from Disney to the Exxon Valdez.* Between the Lines.

Wolf, W. (1996). *Car mania: A critical history of transport* (G. Fagan, Trans). Pluto Press.

Yoos, J., & James, V. (2016a). *Parallel cities: The multilevel metropolis.* Walker Art Center.

Yoos, J., & James, V. (2016b, May). The multilevel metropolis. *Places Journal.* Retreived from https://placesjournal.org/article/multilevel-metropolis-urban-skyways/

Zadražilová, L. (2007). Domov na sídlišti: Mytus nebo realita? [Home in the sídliště: Myth or reality?] In L. Hubatová-Vacková & C. Říha (Eds.), *Husákova 3 + 1: Bytová kutura 70. let* [Husák's apartments: Housing culture in the 1970s] (vol. 3 + 1, pp. 39–56). VŠUP.

Zadražilová, L. (2013). *Když se utopie stane skutečností* [When utopia becomes reality]. Uměleckoprůmyslové Museum in Prague, Arbor vitae.

Žák, L. (1947). *Obytná krajina* [The inhabitable landscape]. SVU Mánes, Svoboda.

Zarecor, K.E. (2011). *Manufacturing a socialist modernity: Housing in Czechoslovakia, 1945–1960.* University of Pittsburgh Press.

Zarecor, K.E. (2018). What was so socialist about the socialist city? Second World urbanity in Europe. *Journal of Urban History, 44*(1), 95–117. https://doi.org/10.1177/0096144217710229

Zusi, P. (2004). The style of the present: Karel Teige on constructivism and poetism. *Representations, 88*(1), 102–24. https://doi.org/10.1525/rep.2004.88.1.102

Zusi, P. (2013). Vanishing points: Walter Benjamin and Karel Teige on the liquidations of Aura. *Modern Language Review, 108*(2), 368–95. https://doi.org/10.5699/modelangrevi.108.2.0368

Index

Page numbers in italics refer to figures.

GLOBAL SUBURBANISMS

Series Editor: Roger Keil, York University

Published to date:

Suburban Governance: A Global View / Edited by Pierre Hamel and Roger Keil (2015)

What's in a Name? Talking about Urban Peripheries / Edited by Richard Harris and Charlotte Vorms (2017)

Old Europe, New Suburbanization? Governance, Land, and Infrastructure in European Suburbanization / Edited by Nicholas A. Phelps (2017)

The Suburban Land Question: A Global Survey / Edited by Richard Harris and Ute Lehrer (2018)

Massive Suburbanization: (Re)Building the Global Periphery / Edited by K. Murat Güney, Roger Keil, and Murat Üçoğlu (2019)

Critical Perspectives on Suburban Infrastructures: Contemporary International Cases / Edited by Pierre Filion and Nina M. Pulver (2019)

The Life of North American Suburbs: Imagined Utopias and Transitional Spaces / Edited by Jan Nijman (2020)

In the Suburbs of History: Modernist Visions of the Urban Periphery / Steven Logan (2021)

www.ingramcontent.com/pod-product-compliance
Lightning Source LLC
Chambersburg PA
CBHW030242030426
42336CB00009B/210

* 9 7 8 1 4 8 7 5 2 5 4 3 9 *